FORBIDDEN
HISTORY
Banned
Maps

Ancient Charts, Ley Lines, and the Geographic
Mysteries That Redefine Our Past.

Ben Wilder

FORBIDDEN HISTORY

Table of Contents

Introduction

The World They Told Us Didn't Exist

How could a map drawn in 1513 show parts of Antarctica centuries before we "discovered" it—apparently without ice? Why do ancient charts sometimes fix longitudes with a confidence sailors supposedly didn't achieve until the 1700s? And what about those maddening curiosities that keep resurfacing like buoys after a storm—phantom islands that haunted official atlases for centuries, and subtle straight-line alignments that seem to link far-flung monuments across continents?

Welcome to *Forbidden Maps*: a journey into the archives, the ocean floor, and the earth itself, where geographic knowledge refuses to stay inside the neat borders drawn by our textbooks. What follows is not a sermon for or against orthodoxy. It's an invitation to look closely—at paper browned by centuries, at coastlines that rise from sonar shadows, at stone raised with improbable precision—and to ask, calmly and relentlessly: what have we overlooked?

Why some maps and geographic traditions don't fit the official story

In 1929, curators in the old imperial palace at Constantinople unrolled a parchment chart dated 1513 and signed by an Ottoman admiral. The western half survives: a sweep of the Atlantic showing Africa and the Americas. What jolted early readers wasn't just that it was an early map of the New World. It was the confidence of its geometry: coasts set in noticeably good longitudes, and hints (depending on who's reading) of a landmass far to the south—an Antarctica-like presence where none should be. The compiler himself boasted he'd drawn on about twenty earlier source maps—some ancient, he said—an audacious claim scholars initially waved away. Yet the chart's consistent longitudes, unusual for that era, forced a second look.

That southern "coast" became the spark for a century of arguments. Some saw in it an outline of ice-free Antarctic shores; others saw a miscopied, distorted extension of South America, a copyist's swerve or a projection's trick. The debate widened when other Renaissance-era maps were reexamined, including a 16th-century world map associated with a European cartographer whose southern continent seems— again, to some eyes—drawn with a familiarity belied by history's official timeline.

If you're already feeling the tug-of-war, good. That tension is where this book lives: between what ink on old vellum appears to say, and what our modern frameworks permit us to hear.

1513 PIRI REIS ATLANTIC, WITH
HYPOTHETICAL SOUTHERN COAST (RECONSTRUCTION)

How forbidden cartography challenges history, science, and religion

This book isn't just about cartography; it's about a pattern of geographic knowledge that falls between the lines of our official story.

- **Ancient Charts:** These are maps that appear to draw on sources older than their date implies—sometimes boasting longitudes before the age of reliable marine chronometers, hinting at coastlines we "shouldn't" have known. An early

1500s compilation using older sources, and a mid-1500s world chart with a striking southern landmass, are prime examples that demand patient attention rather than reflexive dismissal.

- **Ley Lines & Sacred Geography:** Across landscapes, monumental sites—stone rings, pyramids, processional ways—sometimes fall into alignments suspiciously straight over surprising distances. To some, it's pattern-seeking gone wild; to others, it's deliberate sacred geometry and a memory of geodetic arts that bound sky to earth. We'll examine alignments and the astronomical and geomantic thinking behind them, weighing romance against rigor.

- **Suppressed Atlases:** Throughout history, archives in Alexandria, Constantinople, and other centers acted as reservoirs where maps were copied, collated, and sometimes spirited away by war, fire, and politics. The fragments that reach us—survivors of library burnings, seizures, and private hoards—can look like glimpses through a keyhole. When the 1204 crusade redirected its fury to Constantinople, for instance, map collections plausibly shifted west; the survival path of certain charts fits that chaotic relay.

The Term "Forbidden"
"Forbidden" doesn't mean "mystical" or "illegal." It means evidence that never quite makes it into the syllabus: archives that burned, charts copied from older sources we no longer possess, submerged coasts sealed under rising seas, and alignments dismissed as coincidence before they're measured. The prohibition is cultural and methodological—a quiet sidelining more than a door with a lock.

The central problem: the knowledge filter

Why aren't these cases taught as standard puzzles in basic history classes? Because every discipline operates with a **knowledge filter**—a selection process that privileges data that fits established models and quietly quarantines the rest. The filter isn't a conspiracy; it's a survival mechanism. But it has side effects: data that could refine or complicate our picture gets labeled "anomalous" and set aside, sometimes for generations. That's how submerged structures a few kilometers off India's southeast coast, lying in 23 meters of water, can be documented and then go largely unpursued for years—because they raise questions whose timelines feel inconvenient.

The same applies to ancient city-building pushed back earlier than comfortable. Sites like Jericho and other early settlements have been known for decades, challenging the tidy narrative that complex urban life couldn't exist deep in prehistory; yet popular education often treats them as outliers rather than stepping stones. The filter works by insisting that outliers don't rhyme. Our task is to test whether they do.

How Filters Work
1. *Set a model.*
2. *Accept data that supports it.*
3. *Label contradictory data as "noise."*
4. *Forget to revisit the "noise" when tools improve.*
5. *Repeat until a paradigm cracks or a new generation asks impolite questions.*

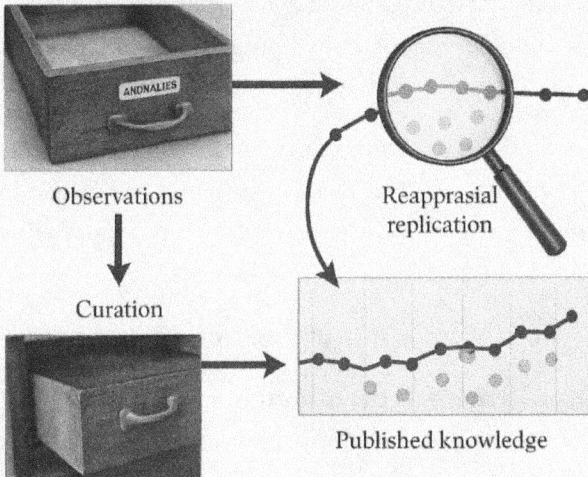

The Knowledge Filter
(a.k.a. file-drawer effect)

Observations

Reapprasial replication

Curation

Published knowledge

Benchmark: concept mirrors the "file drawer problem" in meta-analysis and publication bias flowcharts used in *Cochrane,* Nature articles

Your role: investigator, not believer

I'm not asking you to accept any extraordinary claim on faith. I'm asking you to examine a chain of evidence that includes:

- Early modern charts whose compilers explicitly cite older sources and that display unexpected accuracy in some regions.

- Renaissance and Enlightenment-era maps that appear to "remember" now-icebound or altered coasts, and old atlases that preserve phantom islands which recur across editions for centuries before vanishing from official cartography.

- Underwater structures off India, Japan, and the Mediterranean littoral, logged by national institutes and independent researchers, which complicate assumptions about when and where complex building flourished—especially when sea-level curves are taken seriously.

- Monument alignments and geodetic patterns that may reveal more about prehistoric surveying and sky-earth ritual than our slogans about "primitive" ancestors allow.

We will stress-test these claims. Sometimes the romantic reading will break. Sometimes the prosaic "error in copying" or "projection artifact" will win—and we'll say so. Other times, the anomaly will survive the pressure and become more interesting, not less.

What you will discover in these pages

Part I: Ancient Charts That Shouldn't Exist – We begin in the archives, where early 1500s compilers quietly confess they drew from more ancient maps, and where some mid-1500s world maps sketch a southern land with suspicious specificity. We'll look squarely at the claim that portions of Antarctica were mapped before modern explorers, and at the mainstream counters: misidentified coastlines, copying errors, and projection illusions. We'll also consider the perennial problem of longitude, and whether the observed accuracy in some early charts can be explained by dead reckoning and portolan methods alone.

Part II: Ley Lines & Sacred Geography – We'll step outside the archives and onto the earth, tracing claimed alignments between monuments and decoding the ritual logic of "sacred geography." Expect detours through archaeoastronomy, ritual pathways, and the geometry of place. The aim isn't to prove a mystical grid; it's to examine whether certain builders surveyed with more sky-ground awareness than our stereotypes grant them.

Part III: Suppressed Cartography – Here we'll follow the custody chain of maps: how knowledge was collected in hubs (Alexandria, Constantinople), scattered by catastrophe, and sometimes recopied into new works whose compilers coyly acknowledged their debts. We'll ask what vanished catalogues might have contained—and what the survival of certain motifs implies about older prototypes.

Part IV: What It Means Today – Finally, we'll connect this "forbidden" geography to modern tools—LIDAR, marine geophysics, AI-assisted pattern recognition—and see how new surveys dredge up old debates. Where underwater structures lie at depths consistent with late Ice Age sea-level rises, we'll consider timelines that place sophisticated activity earlier than the standard narrative. We'll also include a **Bonus Workbook** to help you trace alignments, overlay old charts on modern coastlines, and query digital archives for yourself.

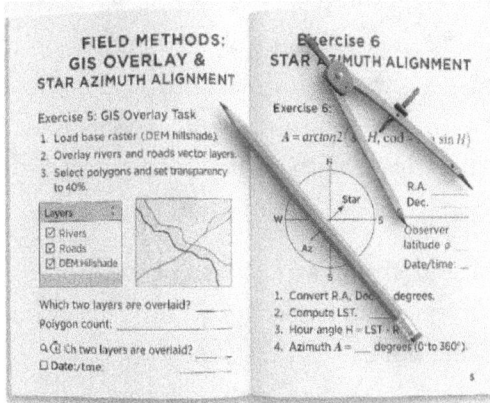

The Bonus Workbook
You'll get practical exercises:

- *How to overlay a Renaissance coastline on modern GIS basemaps.*
- *How to test a claimed monument alignment against random baselines.*
- *How to read bathymetric contours to assess whether a submerged "structure" could be natural.*
- *How to keep a research log that separates evidence from interpretation.*

Raising the stakes: why it matters

This isn't a parlor game for map nerds. The implications are large:

- **Age and reach of civilization:** If even a fraction of the puzzle pieces are authentic—if a few charts truly drew on older, precise surveys; if a few submerged sites prove to be man-made structures built above sea level—then civilization's timeline gets fuzzier at the edges, its early chapters older and more geographically expansive.

- **Navigation and mathematics:** Long before marine chronometers, someone would have needed a practical method—be it astronomical, geometric, or iterative coastal surveying—to fix positions more accurately than expected. That raises engineering and mathematical questions we can and should investigate with today's tools rather than hand-wave away.

- **Religion and myth as carriers of memory:** Traditions that speak of flooded lands, sky gods, and "first cities" might encode cultural memories of real landscapes lost to the sea and technical arts that were ritualized rather than written. Treating them as purely literary can blind us to recoverable data. At the same time, treating them as literal without evidence is no better. The middle path is to test.

- **The politics of knowledge:** Libraries burn. Empires loot. Scholars disagree. And once an academic field sets its guardrails, it takes uncommon patience to drive new evidence through them. The problem isn't bad people; it's institutional inertia. That's why an honest survey of "forbidden maps" must constantly compare extraordinary readings with sober alternatives—and must have the courage to say "we don't know" when we don't.

Ice, Sea, and Time

- *Sea levels were ~60 meters lower around the end of the last Ice Age than today in many regions; local variations apply. Structures now lying 20–30 meters deep could therefore predate the Holocene if they were built on dry land. That's a hypothesis, not a conclusion. Test it with geology, not vibes.*
- *Glacial dynamics are complex. Claims of "ice-free Antarctica" must grapple with regional vs. continental conditions and the difference between coastal shelves and grounded ice. Precision matters.*

Sea Level Since the Last Ice Age: Late Pleistocene–Holocene with Submerged Sites

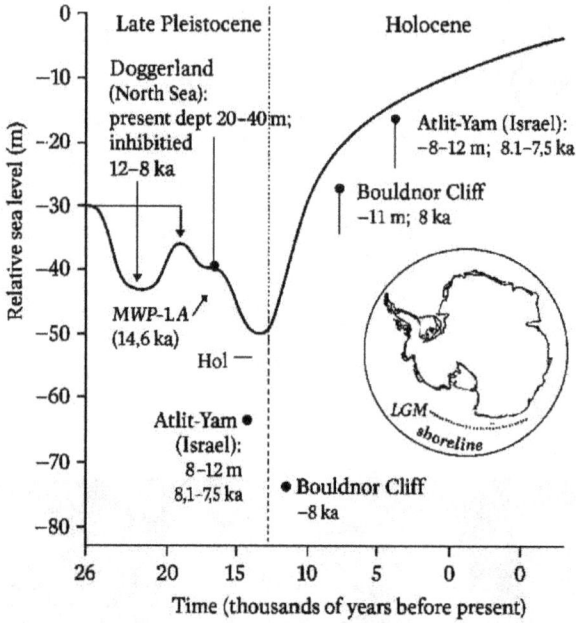

Generalized global curve from Lambeck et al. 2014 PNAS;
MWP-1A atter Stanford 2006, Lin 2021.

A brief tour of the evidence and counter-evidence

The southern continent conundrum. In some mid-16th-century maps, a southern land appears with bays and peninsulas that look—at first glance—uncannily like Antarctica. Those who favor a conservative reading point to errors of projection and the cartographic habit of filling blank space with speculative land. Those who favor a bolder read note features that line up too well to dismiss as luck. We will compare both readings with modern coastlines and ice margins, and ask whether any match exceeds what you'd expect by chance and by the copy-chain of mapmakers borrowing from one another.

Longitude without chronometers? Early modern mariners found latitude easily; longitude was the brutal part. Yet certain historical charts place coasts about each other with a sureness that looks—again, in places—better than guesswork. The conservative explanation is cumulative coastal sailing, compass bearings, and iterative corrections; the adventurous explanation is access to older, more exacting surveys. We'll test segments region by region rather than generalizing from a few striking cases.

Phantom islands and metamorphosing lands. From Antilia to Hy-Brasil to the islands that winked on and off Enlightenment charts, the record shows our ancestors inherited and transmitted geographic memes that took centuries to shake. Rather than laughing at them, we will ask what process created them—distant headlands misplotted, mirages, wishful thinking, or garbled reports of real lands glimpsed in different epochs of sea level. The very endurance of these "ghosts" in atlases is a clue to how knowledge propagates.

Underwater architecture. Professional and independent dives off India's southeast coast documented a U-shaped stone structure at -23 meters depth. Its builders? Date? Unknown. But the depth alone makes it unwise to automatically assign it to recent centuries—sea levels rise slowly on human scales and fast in geological ones. Around Japan, dramatic terraced formations near Yonaguni have split opinion between natural carving and human modification; in the Mediterranean, blocks and harbor works blur the line between ancient coastal engineering and later subsidence. Our method will be consistent: compare depth with local sea-level history; look for unequivocal toolmarks; test for cultural context.

Sacred geography and alignments. Claims about straight lines connecting ancient sites can be made to appear or vanish depending on selection criteria. We'll therefore do what any honest investigation must: set rules in advance (what counts as a "site," what tolerance for deviation), run statistical baselines, and ask whether any alignment survives that level of discipline. Where alignments do hold, we'll explore whether they reflect practical surveying, processional routes, or a cosmology that "wired" land to sky.

Why some maps and traditions don't fit the official story

Every time you see a clean historical narrative—a rising line from "hunter-gatherers" to "cities" to "empires"—remember how much of the curve we've reconstructed from fragments. In the real world, knowledge travels in fits and starts. It leaps across cultures by trade, marriage, war, and religious mission; it hides in monasteries; it drowns with ships. When you take that seriously, odd survivals stop looking like magical anomalies and start looking like what they probably are: **memory shards**—technical, geographic, and ritual—embedded in later works.

- The claim that pre-modern chartmakers compiled older sources isn't romantic; it's what they sometimes wrote in their own marginal notes: "this coast from an ancient source," "that island from the chart of X." Such notes remind us that they were curators as much as authors.

- The notion that certain coasts were seen before official "discovery" dates is not heresy; it's a reminder that sailors' knowledge often precedes imperial fanfares by generations. The

---

The Rules We'll Use
- *No cherry-picking: we'll publish inclusion criteria for sites and lines.*
- *No single smoking gun: we look for patterns across independent domains (maps, geology, archaeology, sky).*
- *No paranormal escape hatches: if unknown, we say "unknown."*
- *No ridicule as argument: if a claim fails, we say why—with measurements.*

---

question is not whether pre-expedition sailors reached certain latitudes; it's how far and with what fidelity they recorded what they saw.

- Traditions of sunken lands in the Indian Ocean or the Mediterranean needn't be taken literally to be useful; they can point to shelf areas that were dry during lower sea levels— prime zones for early settlements that later drowned and silted over.

Memory Shards

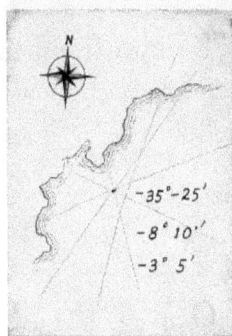

N

$-35°-25'$
$-8°10'$
$-3°5'$

Marginal map notes

May

In view of another country beyond the waters.

Logbook

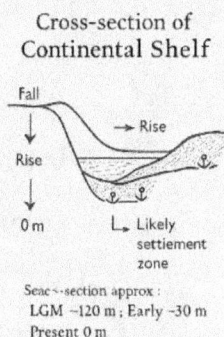

Cross-section of Continental Shelf

Fall

→ Rise

Rise

0 m

L→ Likely settlement zone

Sea section approx :
LGM ~120 m ; Early ~30 m
Present 0 m

Cross-section

How forbidden cartography challenges history, science, and religion

History. A robust history welcomes puzzles. If parts of our early modern map record derive from now-lost surveys—whether ancient Mediterranean, Indian Oceanic, or otherwise—that would complicate our model of knowledge transfer. It would not vaporize history's foundations; it would give them more interesting roots. That, in turn, would force us to re-plot chapters on migration, trade, and exploration, and to consider that precocious maritime cultures may have reached farther, earlier, and more methodically than currently credited.

Science. The scientific method is not allergic to anomalies; it thrives on them when handled exactly: measure, model, re-measure. Ice and sea-level reconstructions can be compared to claimed ancient coastlines. Geophysical surveys can test whether underwater features are natural or architectural. Archaeoastronomical models can test alignments against chance. Where claims fail, the method has done its job; where they survive, we've learned something new.

Religion and myth. Sacred narratives encode cosmology and memory. Some speak of floods, giants, sky travelers, and "first times." Interpreting these literally or allegorically is not the only option. There's a third: **semiotic archaeology**—treating stories as curated containers for older observations about sky cycles, earth rhythms, and cultural trauma (like sudden inundations). That approach neither mocks belief nor suspends skepticism. It reads carefully.

What you will discover in these pages

By the time you reach the end of this book, you will have:

- **Handled the evidence yourself.** You'll have overlaid early charts on modern coasts, noted where they fail and where they spookily hold. You'll have learned how projection choices deform shapes, and how to guard against being fooled by good fits born of bad math.

Three Lenses to Wear
- *Historical lens: custody chains, scribal habits, projection systems.*
- *Scientific lens: geology, bathymetry, statistics, error bars.*
- *Mythic lens: symbols as memory devices, not casual metaphors.*

- **Looked beneath the waves.** With sea-level curves in hand, you'll have assessed claims about drowned cities. You'll know why a structure at 23 meters depth raises a different set of questions than one at two meters—and what kinds of proof (toolmarks, layout logic, cultural artifacts) convert "intriguing" into "compelling."

- **Tested sacred geography.** You'll have run simple alignment tests that any skeptic—or believer—can replicate, and you'll have seen where straight-line romance yields to the gritty joy of measurements and tolerances.

- **Distinguished memory from myth.** You'll have a method for scanning ancient narratives not for miracles but for **signals**—periodicities, sky events, and flood recollections—that can be cross-checked against ice cores, precession cycles, and bathymetric realities.

- **Learned to live with uncertainty.** Some questions won't resolve neatly. That's not failure; that's a more honest map of our ignorance—and a better compass for future exploration.

A word on balance

It's easy to drive into the weeds in this subject. One ditch is credulity: seeing precision where there is none, insisting on ice-free poles without supporting glaciology, and mistaking folklore for field notes. The other ditch is **performative skepticism**: laughing at anomalies instead of measuring them, or refusing to revisit old questions with new tools because the questions embarrass us. This book steers the middle track. When an extraordinary claim fails under scrutiny, I'll say so plainly. But I won't pretend failures discredit the entire inquiry—especially when other claims, more modest and testable, remain.

And yes, some voices in this conversation quickly jump to non-human intelligences in our distant past. While this volume's focus is earthly—maps, monuments, coastlines—we won't banish that discussion. We'll bracket it. If, and only if, the terrestrial evidence forces questions that known human pathways cannot answer, we'll say, "This is where the puzzle widens." Until then, we'll keep our feet on the ground and our eyes on the documents.

A few emblematic puzzles we'll take apart together

The southern "memory" in early modern maps. We'll explore how a 16th-century compiler could plausibly have stitched together older coasting surveys to produce a southern outline that accidentally resembles Antarctica, and what it would take—in terms of sailing seasons, currents, and hull tech—for someone to have charted Antarctic coasts in any detail pre-modernity. We'll weigh those logistics alongside the cartographic evidence.

Metamorphoses of Antilia and other phantom islands. You'll see how a phantom can migrate across atlases for centuries, why "erasing" it took so long, and how sometimes the ghost is the scar left by an older, real island glimpsed during a different sea-level regime or misplotted by a tired pilot.

The longitudes puzzle. We'll build a small "paper lab" to test how portolan-style iterative sailing, compass correction, and star fixes could have produced surprising accuracy in some sectors while failing in others—no mysterious chronometers required. We'll then ask: Does anything remain unexplained after you generously model navigator skill?

Shorelines under the sea. We'll examine that U-shaped offshore structure, the terraced formations of the western Pacific, and Mediterranean harbor remains. For each, you'll see how to differentiate tool-shaped from wave-shaped, human plan from cliff break, and cultural assemblage from random stones. Depth will be our metronome; geology, our drum.

Earth's ritual geometry. With a handful of famous alignments in hand, we'll run distance and azimuth checks and then look for cultural context: were these lines processional, astronomical, or purely modern projections? Some will falter; others may surprise you with their sober logic.

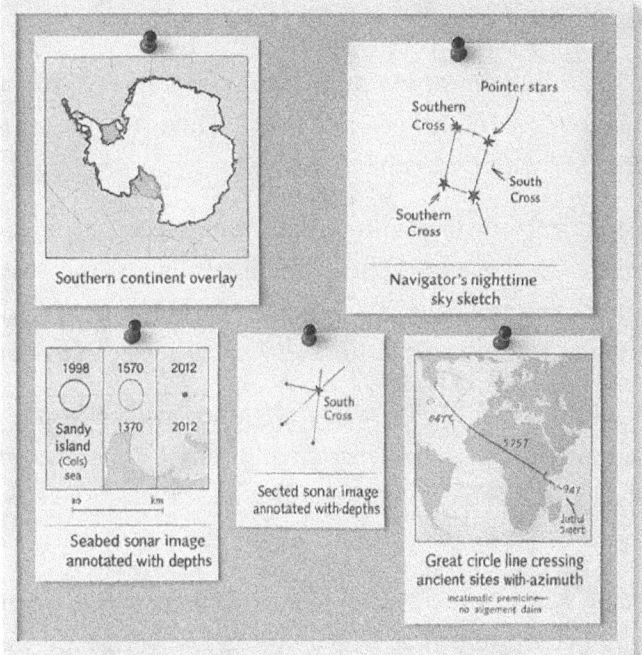

Southern continent overlay

Navigator's nighttime sky sketch

Seabed sonar image annotated with depths

Sected sonar image annotated with depths

Great circle line crossing ancient sites with azimuth

What Counts as a Win

- *Ruling out a seductive but wrong idea is a win.*
- *Finding that a romantic claim survives sober testing is a bigger win.*
- *Discovering we need better data is the best kind of frustration.*

A preview of the emotional terrain

Expect whiplash. On one page, you'll feel wonder—the sense that we've underestimated our ancestors again. The next, you'll feel deflation—"oh, that bay is just a projection artifact." This is healthy. We're training our eyes to hold both skepticism and curiosity without letting either eat the other.

- You'll feel awe staring at an early chart whose coastal sweeps match modern outlines eerily well. Then you'll learn to see the places where it falls apart—and why.

- You'll feel a thrill reading of drowned precincts and stonework under the waves. Then you'll sit with the painstaking questions: toolmarks? Context? Corroboration? And you'll learn to love those questions, because they keep us honest.

- You'll feel the tug of sacred alignments; then you'll calculate azimuths and realize how many "lines" dissolve when you change the inclusion rules. The lines that survive will mean more.

If these maps, alignments, and submerged stones are merely curiosities, then this is a charming tour. But if even a few withstand disciplined scrutiny, the consequences ripple outward. Trade routes extend. The apprenticeship of navigation lengthens. Rituals reveal geometry. Flood myths gain footnotes. And the lost libraries of humanity—burned, looted, or drowned—feel a little less lost.

In short: The maps are real. The anomalies are undeniable. The question is not whether they exist—but why we were never meant to see them.

You're about to step into a world where every line drawn and every symbol placed may hold the key to a lost chapter of human history. This is not a call to blind belief. It's an invitation to disciplined wonder.

Shall we turn the page?

Part I: Maps from a Forgotten World
Chapter 1: The Piri Reis Map

On a winter morning in Constantinople, 1929, a rolled gazelle-skin was uncurled in the former sultan's palace, and a lost world seemed to tilt into view. Painted in jewel tones and salted with marginal notes in Ottoman Turkish, the chart bore a date—Hijri 919, our 1513—and the signature of an Ottoman admiral with the sea in his name: Piri Reis. If you trace the painted coasts with your fingertip, Africa slides by, the bulge clean and familiar; South America leans away, overlong and sinuous; and then—down in the map's austral margins—there is an insinuation of something else. An ice-shaped absence that some readers swore was no absence at all but a presence: the shore of a southern continent that would not be "discovered" for three more centuries.

That is the spark that ignited the modern mystery. What, precisely, are we looking at?

A map that talks back

This map is unusually talkative. Along its coasts, Piri Reis wrote short notes explaining what he drew and where his information came from. He said he compiled it from "about twenty old charts and eight mappaemundi," including documents "prepared at the time of Alexander," four Portuguese world charts with mathematical construction, and a chart of the West Indies that he credited to Columbus. He emphasized that he "put all these together on a common scale."

Those notes matter. They tell us the 1513 parchment is a copy-and-blend of many earlier sources—some recent, some older, some said to be very old. They also remind us that the surviving piece is just that: a fragment of a larger world map whose missing half might have included additional coasts.

Now, fixed to the page, the puzzle deepens. On many read-throughs by modern researchers, the southernmost coastline on the Piri Reis fragment appears to mirror the profile of Antarctica's Queen Maud Land—not as today's ice-sheathed scarp, but as a crenulated, river-notched littoral, as if surveyed when water still ran off its mountains to the sea. The claim is extraordinary. But this is a chapter about weighing the extraordinary—open-minded, evidence-first, and constantly checking ourselves against what we know and what we merely want to be true.

A 16th-century chart showing Antarctica without ice

If you only glance, the "Antarctic" claim sounds like myth-making. But several converging observations built the case that something unusual is recorded on the Piri Reis parchment and kindred Renaissance maps:

1. **Ice-free profiles.** A Cold War–era technical review noted the Piri Reis coastline matches the **subglacial** profile—the bedrock outline hidden beneath the Antarctic ice—mapped only in the mid-20th century by seismic surveys. That letter's dry phrasing ("we have no idea how the data on this map can be reconciled with the supposed state of geographical knowledge in 1513") is part of why it is so often quoted. The provocative implication: the source map behind Piri Reis was made when that coast was *not* under an ice shell, or by someone with knowledge of the bedrock beneath it.

2. **A family of southern maps.** Piri Reis is not alone. A 1531 world map by the French cartographer Oronteus Finaeus sketches a southern landmass ringed with mountains, river systems flowing to the sea, and a south pole placed plausibly near the center. Later analysis compared these features to modern seismic "bedrock" charts and found surprising congruities—especially around the Ross Sea, where Oronteus draws fjordlike inlets where we now see outlet glaciers.

3. **A chronology that could fit.** If (emphasis on *if*) those southern coasts are ice-free depictions, when could such a survey have happened? The argument advanced by several investigators runs like this: large sectors of Antarctica's periphery might have been seasonally or persistently less ice-choked late in the last Ice Age and during the early Holocene, with some coasts plausibly open until about 4000 BC, after which the advancing ice-shelf sealed them. In that scenario, the older source for Oronteus reflects an earlier, less glaciated moment, while Piri Reis inherited coastlines from later, more ice-encumbered but still open shores.

None of that proves a lost civilization. It does, however, justify a deeper look at what Piri Reis actually compiled, how he said he compiled it, and why both mainstream and alternative readings persist.

How it suggests ancient seafaring beyond known history

First, a sense of genre. The Piri Reis fragment is drawn like a **portolan**—a mariner's chart whose backbone is a compass-rose lattice. Coastlines are emphasized; interiors are ignored. Colors speak a code: rocky black dots, sandy red stipples, fortified ports in red, all using a visual language Piri Reis says he consciously adopted from the "international cartographical traditions" of his time.

He also tells us exactly what he did: he scaled disparate sources into a single framework and stitched them. That method can create systematic distortions—overlaps, double-drawn rivers, elongations—especially when the source charts use different projections or are themselves copies of copies.

Example: the Amazon appears *twice* on the Piri Reis chart, most plausibly because two source documents overlapped the same region at different scales. One depiction seems to show the river reaching the Atlantic without Marajó Island marked; the other shows Marajó in notable detail, centuries before Europeans officially recorded it. That does not demand a prehistoric survey; it does demand that Piri Reis had access to surprisingly precise information he explicitly said he didn't personally observe.

Example: the Orinoco, as we know it, is not portrayed. Instead, the chart shows two broad estuaries cut far inland at roughly the right latitude/longitude—suggesting either delta progradation since the source map was drawn or, more prosaically, source-map mismatches in scale and shoreline detail. Both are possible; neither is trivial.

What the Admiral Wrote

- *"I made use of about twenty old charts and eight mappaemundi... of charts prepared at the time of Alexander... of four Portuguese charts and the chart Columbus drew for the West."*
- *"Putting all these together on a common scale, I produced the present map."*
- *"My map is as correct and dependable for the seven seas as the charts that represent the seas of our countries."*
These are the map's claims about its sources and method. Treat them neither as gospel nor as fiction; treat them as data.

And that southern coast? If you interpret the map "literally," the profile seems to echo Queen Maud Land's bedrock outline, not its cliff of ice. If you interpret it "skeptically," the same profile can be read as a stylized extension of South America—the common Renaissance habit of inflating Terra Australis to "balance" the globe, with no claim to real survey. The truth may be a third thing: a spliced coast that accidentally aligns with modern subglacial outlines because Piri Reis was remarkably faithful to some older coastal geometry…without knowing what he had.

حليه كراي
عدكلر

PİRİ REİS
MAP (1513)

A detective story in three acts

Act I — The Rediscovery (1929)

The map did not loom large in sixteenth-century Europe. It flared into modern awareness only when Turkish archivists found it in the old imperial palace in 1929. That rediscovery, plus early press coverage, drew both scholarly curiosity and nationalist pride—this was, after all, a very early world chart made by a mariner of the Ottoman world.

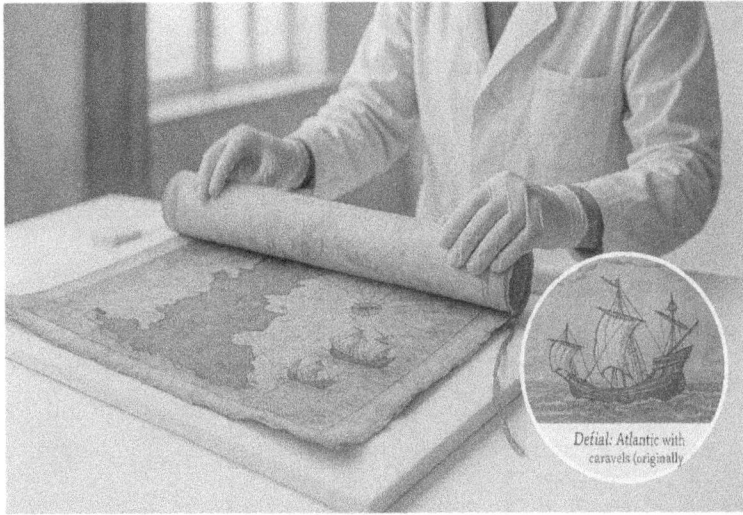

Defial: Atlantic with
caravels (originally

Act II — The Hypothesis (mid-20th century)

Mid-century analysts with access to newly mapped Antarctic bedforms looked south on the Piri Reis and Oronteus Finaeus charts and asked: coincidence, or continuity? When the 1949 seismic profiles of Queen Maud Land's subglacial coastlines were compared to the map's shapes, the odd parallels were logged. When the 1531 Oronteus map showed river systems where modern ice streams debouch, eyebrows rose again. And when multiple Renaissance charts seemed to carry cognate southern geometries, a bolder hypothesis congealed: the Renaissance copies were digesting fragments of far older surveys.

Act III — The Counter-Case (late-20th to 21st century)

Cartographic historians and glaciologists countered with strong medicine: Renaissance "southern continents" were almost always conjectural; portolan-style stitching creates phantom capes and ghost rivers; projection mismatches can make unlike things look alike. On this reading, Piri Reis's austral fringe is not "Antarctica without ice" but a distorted South America and speculative Terra Australis that *happens* to mimic parts of a bedrock coast—an illusion born of stitched sources and a human eye hungry for pattern.

In short, two coherent readings stare at the same ink: one that spots geological signal; another that sees cartographic noise. The honest way forward is to test specifics.

How could it be true — a balanced scenario

Suppose for a moment that the southern coast on the Piri Reis fragment *does* echo a real, ice-light shoreline. How do we tell a story that does not require time travel or clairvoyance?

- **Transmission chains are real.** Knowledge traffics along long cultural routes. Piri Reis himself sketches a plausible pipeline: classical-era geographic tables, Hellenistic mappaemundi, Islamic mathematical charts, Iberian pilot books, and a Columbus chart. He suggests that compilations of older charts were studied in the great libraries of the Mediterranean, particularly Alexandria and, later, Constantinople. Fragments could have survived the destruction of one center by hiding in the archives of another.

Three Things the Piri Reis Map Is

- *A compilation. It is not an eyewitness survey. It is a cleaned-up collage. Piri Reis says so.*
- *A data hoard. It visibly preserves coastal details (e.g., in the Amazon estuary system) that exceed what most 1513 Europeans could access.*
- *A fragment. We lack the missing half. Conclusions based on the surviving shard must be provisional.*

- **Maritime techniques can be better than we think.** We chronically underestimate what pre-modern navigators could do with dead reckoning, coastal piloting, astronomical bearings, and a shared tradition of rhumb-line sailing recorded on portolans. Even without precision chronometers (an eighteenth-century invention), coastlines can be traced remarkably well, especially if surveyed repeatedly over generations. A chain of cultures could, collectively, improve a coastal picture far beyond what any one expedition could achieve.

- **Climate windows existed.** Most of Antarctica has been ice-capped for a very long time, but the thickness, extent, and shelf margins have fluctuated. The "window" view is that some coastal sectors may have been more open in the late Pleistocene/early Holocene, then became progressively sealed by shelves and grounded ice thereafter. If true, a patient, multi-epoch tradition of chart copying could preserve a vestigial outline of those former coasts.

This is a *possible* story, not a proven one. But it is not an unreasonable one—and it respects both the map's testimony and the caution demanded by glaciology.

How could it be wrong

Much of the skepticism comes down to three sober points:

- **Projection illusions.** Stitching sources drawn on different projections creates lobes, kinks, and elongations that hindsight can mistake for real headlands. The "Antarctic fit" may be a projection artifact of our pattern-hunger completes.

- **Terra Australis syndrome.** For centuries, European mapmakers "needed" a southern land to balance the globe; the habit of drawing a notional continent was pervasive. In that milieu, Piri Reis's southern smear could be conventional cartographic ballast, not archaic memory.

- **Selective confirmation.** Once you're primed to look for an Antarctic fit, you may under-weight the misfits (scale errors, wrong peninsula length, capes that go nowhere) and over-weight suggestive overlaps. Oronteus's rivers, where our glaciers are, may be the most impressive overlay—but even there, the differences are non-trivial.

A balanced chapter admits as much. The question is not "is the Piri Reis map *definitely* Antarctica?" but "does the cumulative weirdness warrant keeping the file open?" I argue it does.

The Five Most Testable Claims

1. The austral coastline on the Piri Reis fragment matches Queen Maud Land's **subglacial** profile better than it matches a stylized South America. Test: quantitative outline comparison with modern bedrock contours.

2. Oronteus Finaeus's Ross Sea rivers align with the present-day positions of major outlet glaciers. Test: fjord/glacier mouth proximity analysis.

3. The Amazon double-rendering implies overlapping sources of different ages and scales. Test: reconstructing likely pilot-chart lineages that would produce the duplication.

4. The estuary forms near the Orinoco's latitude indicate a markedly different delta regime at the time of the source chart. Test: stratigraphic/deltaic progradation models vs. early modern shoreline sketches.

5. The map's projection framework can be reverse-engineered from its rhumb-line lattice and coast geometry. Test: fit the chart to candidate projections and quantify distortions.

Case Study: Charles Hapgood and the theory of Ice Age cartographers

The most ambitious interpretation proposes that the Renaissance mapmakers were digesting **copies of copies** from a seafaring culture, or network of cultures, capable of sustained coastal survey at the end of the last Ice Age. In this telling, skilled navigators charted long sections of shore while the world was re-shaping: seas rising, shelves flooding, river mouths migrating. These charts—perhaps kept as maritime trade secrets—were archived in ancient centers of learning, excerpted by later compilers, and finally distilled into the early modern atlases we can hold.

This "Ice-Age cartographer" hypothesis lives or dies on three pillars:

1. **Technical plausibility.** Could pre-Neolithic mariners really do this? For **coastal** survey, yes. With repeated soundings, lead-line measurements, careful bearings, and a standardized way to draw, a tradition of pilotages can approach an astonishing fidelity along shore. The missing piece is *longitude*—but a chart can be locally accurate without modern longitudes if it evolved segment-by-segment.

2. **Transmission plausibility.** The Mediterranean and Near East are history's data superhighway. It is reasonable that a corpus of old pilot charts could have moved from port archives to libraries, been copied into mappaemundi and portolans, and then into early modern compilations. Piri Reis explicitly names

this stew of sources and says he synchronized them on a shared scale.

3. **Geological plausibility.** The end of the Ice Age was not a single flood but a stair-step of meltwater pulses over millennia. If certain Antarctic fringes were indeed seasonally more open before ~4000 BC, a determined maritime culture might have mapped discrete sectors (especially peninsulas and embayments), leaving behind a patchwork that later compilers inelegantly merged.

Is this proven? No. Is it falsifiable? In principle, yes: if every specific "fit" dissolves under quantitative comparison or can be better explained as projection artifacts, the hypothesis fails. If even one sector holds up under rigorous overlay, the hypothesis stays interesting.

A Researcher's Reading List (What to Seek, Not Whom to Believe)

- *Primary artifact: high-resolution facsimiles of the 1513 Piri Reis fragment and the 1531 Oronteus Finaeus world map; examine the marginalia and rhumb lattices.*
- *Geophysical comparators: bedrock contour maps of Queen Maud Land and Ross Sea sectors from mid-20th-century seismic campaigns; overlay carefully at comparable scales.*
- *Transmission context: records of Constantinople's imperial library holdings; medieval portolan traditions linking Islamic, Iberian, and Italian pilots; the practice of compiling "world" charts from regional pilots.*

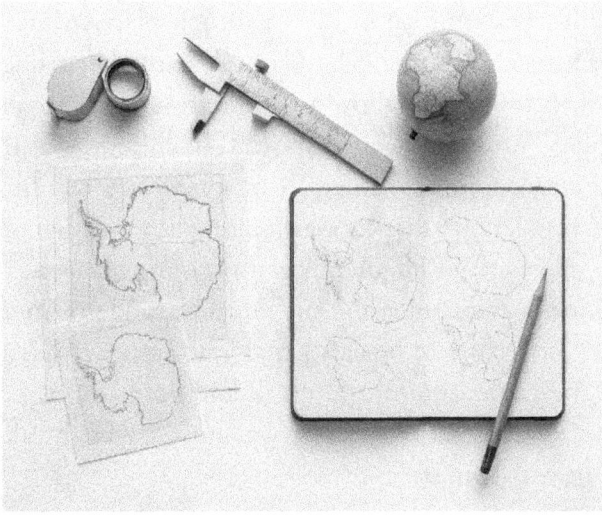

A guided tour of the 1513 coast

Let's "walk" the Piri Reis fragment as a mariner would, left to right:

- **The African bulge** is crisp; West Africa's contour aligns closely with modern charts—unsurprising given Portuguese and Ottoman sailing competence on that route in 1513.

- **The Brazilian "arc"** is elongated and flows too far south— likely a scale issue in the underlying source, or a blend of two sources with different bends.

- **The southern swirl.** Below South America, the coastline that stirs this entire debate sweeps eastward in lobes and bays that, under one reading, resonate with Queen Maud Land's bedrock profile. Under another, they are merely South America's tail stretched by a bad projection and a cartographer's urge to "close the globe" with a southern counterweight. The same evidence, two plausible readings.

The Library Moment

A particularly vivid episode in this saga took place in the Library of Congress during the 1959–60 holidays. A researcher requested every old world map that might show a "southern land." Dozens of charts were laid out. Turning a manuscript leaf, he found himself staring at Oronteus Finaeus's 1531 southern hemisphere and felt, in his words, "an instant conviction" that he was looking at an authentic southern continent—not fantasy. He then began measuring traverses across the drawn continent and comparing the ratios to modern Antarctica; the figures were "too close to be accidental," he argued. Whether or not you agree, the scene captures the thrill and peril of this field: in old lines, our eyes find patterns; then we must try to break our own enchantment with numbers.

Why this chapter belongs in a book called *Forbidden Maps*

Because "forbidden" is often a feeling, not a fact. It's the feeling one gets when evidence doesn't fit the syllabus, when marginalia in a sixteenth-century hand threaten to upstage neat timelines. The Piri Reis map forces us to hold two ideas at once:

- Human cartography is conservative and cumulative; maps are palimpsests of shared practice.

- Human prehistory is more dynamic than our diagrams; seas rise, coasts drown, libraries burn, and knowledge bottlenecks.

If you want only certainties, old charts will frustrate you. If you relish difficult maybes, they will sing.

What, then, should we say of the Piri Reis map?

We should say it is a masterpiece of **compilation**: an Ottoman admiral frankly described his method; he had privileged access to archives; he handled multiple, uneven sources; and he scaled them into a single sea-book chart. In the process, he preserved coastal information that reads as uncanny from a European 1513 vantage—uncanny, but not impossible, given the data networks of the Mediterranean world.

We should say its austral coast is **interpretable** in two coherent ways: as a stylized extension of South America/terra australis, or as a shadow of an older coastal geometry that aligns in places with Antarctica's hidden bedrock coast. Both interpretations have strengths and weaknesses; both deserve to be plotted, measured, and argued in good faith.

We should say that a family of early-modern maps—Piri Reis, Oronteus Finaeus, Mercator strands—collectively **raises a question**: did geometric knowledge of parts of the far south survive in Europe from older, perhaps ancient, source charts? The Ross Sea rivers on Oronteus are the most tantalizing datum; they are also the ones crying out for rigorous, reproducible tests.

And finally, we should say that whether or not these maps prove an Ice-Age surveyor, they **prove** something else that matters for this book: the world's memory is not a straight line. It is a tide that advances, retreats, and leaves shells of knowledge on unexpected shores.

Takeaways

- *Never trust a single overlay. Always test multiple scale fits and projections.*
- *Read the marginalia. The map tells you how it was made. Believe it… But verify it.*
- *Beware of "it looks like." Pattern-matching is a beginning, not a conclusion.*
- *Follow the chain. The question isn't "who first drew Antarctica?" but "how did specific coastal geometries survive into 1513?"*

Chapter 2

The Oronteus Finaeus and Buache Maps

You're about to stand in front of a wall-sized atlas no one was meant to inherit. Dim parchment. A hemisphere traced in confident lines. A stark white emptiness at the bottom of the world... except it isn't empty. It has rivers. Capes. Bays. Mountain backbones that seem to breathe beneath the ink.

What you're looking at—what we will unpack together—is one of the strangest crossovers in the history of knowledge: early-modern maps that appear to draw the polar regions with a nuance their makers shouldn't have possessed—and the new, hidden Antarctica we're discovering today under two to four kilometers of ice. This chapter makes that bridge tangible and testable.

I'll guide you through two "forbidden maps"—the 1531 world map of **Oronteus Finaeus** and the mid-18th-century **Buache** map—and set them against what satellites, ice-penetrating radar, and physics-based reconstructions now let us see beneath the Antarctic Ice Sheet. We'll keep our minds open and our standards high: balancing mystery with measurement; weighing bold hypotheses against what the data actually say. You'll finish this chapter not as a passive reader but as an investigator who can trace the evidence line by line.

A Map that Shouldn't Be There (and Yet Is)

In the winter of 1959–60, a researcher leafing through hundreds of old charts in Washington, D.C. froze mid-turn. On the southern half of a world map dated 1531, a continent sat where the "Unknown South Land" usually loomed as pure fantasy. But this one—penned by the French mathematician-cartographer Oronce Fine (Latinized: Oronteus Finaeus)—showed something different: a southern land with specific features—coastlines, mountain chains, and what looked like natural **river drainage**—and a polar point not wildly misplaced. The close look that followed argued this was no mere flourish of Renaissance imagination; the coastlines were individualized, the rivers drained as rivers do, and the interior—strikingly—was left largely blank, as if the ice had kept the heart of the continent inscrutable while its edges remained open water.

When this 1531 Antarctica is compared, proportion for proportion, to modern outlines (and adjusted for the period's quirky projections), several coastal stretches and relative distances fall into uncanny corridors of agreement. Analysts have argued the **scale is off**—the southern land balloons too far toward the tropics, Drake Passage is pinched—but the **patterning** of the coasts and the logic of the rivers have kept the conversation alive for decades.

What to Notice First on the Oronteus Finaeus Map

- *Coastal mountains that run roughly where ranges ring Antarctica today.*
- *Deep inlets where modern ice streams now debouch into the sea.*
- *River-like tracings flowing seaward, suggesting ice-free littorals at the time of the source map(s).*
- *A conspicuously blank interior—possibly reflecting ice-bound highlands rather than ignorance.*

Terra Australis on Oronce Finé's
Double-Cordiform Map (1531) - with
modern Antarctica outline (gost) for comparison

The Buache Shock

Jump forward two centuries. In the 1730s–1760s, **Philippe Buache,** a Paris geographer and royal academician, published maps of the Southern Ocean and a speculative **Carte des Terres Australes.** Unlike many armchair fantasies of Terra Australis, Buache's rendering of the far south included an **inland polar sea** (he called it the *Mer Glaciale*) and a **through-continent waterway** dividing the south polar lands into eastern and western masses. The map reads, on its face, as a conceptual synthesis—a physical geography designed from basic principles (mountain watersheds, iceberg calving requirements) plus traveler reports from high southern latitudes. But it happens to **resemble** something we would only verify in outline **two centuries later**: that beneath the ice, large basins and troughs carve across Antarctica; if stripped of ice, the "continent" would resolve into a complex **archipelago** bounded by shallow seas and linked basins.

Buache's hypothesis of a divided Antarctica has often been dismissed as pure conjecture. Indeed, historians note it was an **inference** supported by then-current reports of enormous icebergs (which Buache reasoned must calve from vast interior waters fed by "great rivers" and mountain chains), not the copying of any ancient source map. Still, the **form** of his guess—a trans-Antarctic waterway and submerged basins—maps eerily onto what the best **bed topography** models show today.

Buache's "Mer Glaciale": Hypothesis or Memory?

Buache presented his polar sea to the Académie des Sciences as a logical reconstruction: mountains beget rivers; rivers feed an inland frozen sea; such a sea calved the titanic bergs explorers reported. Whether he drew on lost source maps or not, his design anticipated the idea that Antarctica's "continent" is, in bedrock terms, not a solid block.

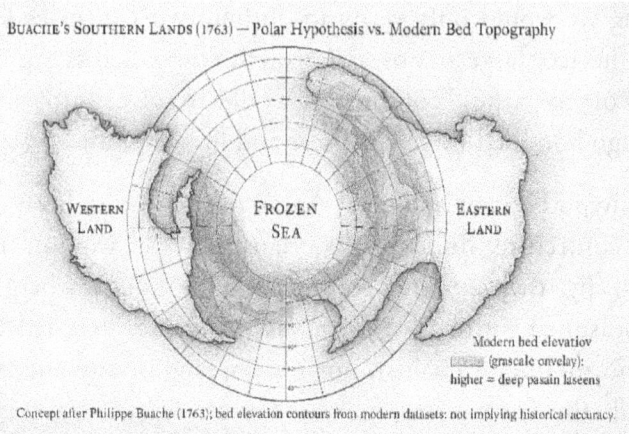

BUACHE'S SOUTHERN LANDS (1763) — Polar Hypothesis vs. Modern Bed Topography

WESTERN LAND

FROZEN SEA

EASTERN LAND

Modern bed elevatiov (grascale onvelay): higher = deep pasain laseens

Concept after Philippe Buache (1763); bed elevation contours from modern datasets: not implying historical accuracy.

Ancient knowledge of Polar Regions

How could 16th- and 18th-century mapmakers know anything credible about regions no ship had charted? The orthodox answer is that they **didn't**: Terra Australis was a balancing act—Europeans assumed there must be southern land to counterweight the known continents. But the **details** on a few maps beg for careful audit. The Finaeus map, for instance, was compiled from **multiple sources**, each on different projections, which can explain distortions—yet the **coastal drainage** motifs and the "feel" of the drawn terrain are hard to dismiss as mere decorative flourish. Several researchers have argued that, when corrected for projection and scale, **segments** of the Finaeus coastline line up with parts of East Antarctica and the Ross Sea margins; and where modern science shows **no rock shore at all** (just basins lying below sea level under the ice), the Finaeus depiction **goes vague**— which is precisely what you'd expect if a coastal shelf were indeed submerged.

The possibility raised in many "forbidden map" debates is not that Renaissance cartographers sailed Antarctic beaches, but that they sometimes compiled **older charts**—fragments or derivative coast-sketches—from libraries whose collections were wiped out or scattered

(Alexandria, Constantinople), preserving a **memory layer** of coastlines from earlier epochs. You don't have to **believe** that hypothesis to see why these two maps won't quite go away: every time modern glaciology peels back the ice with new instruments, **another piece** of Antarctica looks a little more like the hidden land Buache inferred and Finaeus inked.

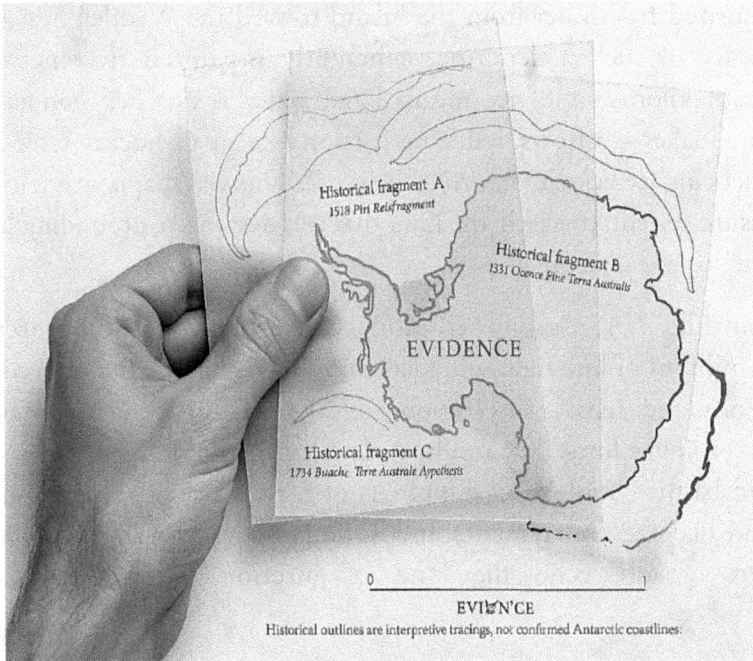

Historical fragment A
1518 Piri Reisfragment

Historical fragment B
1331 Oconce Fine Terra Australis

EVIDENCE

Historical fragment C
1734 Buache Terre Australe Avpothesis

0

EVILꞍN'CE
Historical outlines are interpretive tracings, not confirmed Antarctic coastlines.

River systems under the Antarctic ice sheet

For centuries, we pictured Antarctica's base as a frozen anvil. It isn't. New radar, satellite motion tracking, and mass-conservation modeling have built a radically different picture: a **wet** underside, with hundreds of subglacial lakes and **river systems** that channel water across continental-scale distances to the sea. Over 400 subglacial lakes have been cataloged, with giants like **Lake Vostok** the size of a small country. These lakes and the channels between them act as a **plumbing**

system that switches on and off, flushing water toward the margins and affecting how fast the ice above flows.

In 2022, a team using aircraft-borne ice-penetrating radar and modeling announced something even more startling: a ~460-kilometer subglacial river—longer than the Thames—draining a catchment "as large as Germany and France combined," moving pressurized freshwater from the inland toward the Weddell Sea. The presence of such a dendritic, **coherently organized** river network beneath kilometers of ice means basal water is not just ponded in isolated lakes—it **flows**, reducing friction like an air-hockey table and influencing ice velocity far from the coast. This was mapped as a **high-pressure** system, precisely the kind that can destabilize grounding lines where ice meets ocean.

Meanwhile, NASA-supported products like **BedMachine Antarctica** fuse millions of line-miles of radar soundings with satellite-derived ice motions and mass conservation to calculate the **bed topography** between flight lines. The result? A detailed map of hills, troughs, and broad basins—some lying hundreds of meters below sea level—that would become **embayments and sounds** if the ice vanished. Those basins broadly echo the kind of **interior waterways** Buache

---

The Invisible Rivers of Antarctica (Quick Facts)

- *>400 subglacial lakes identified under the Antarctic Ice Sheet.*
- *A dendritic river system ~460 km long drains a vast inland basin toward the Weddell Sea (2022).*
- *Bed topography reveals multiple below-sea-level basins; without ice, Antarctica becomes a chain of islands and shallow seas.*

hypothesized and the **estuarine** hints sketched on Finaeus's southern shores.

When Rivers in Ink Meet Rivers Under Ice

Let's line the claims up:

1. **Finaeus (1531)** shows coastal Antarctica with **estuaries and rivers** entering the Ross Sea and other margins; the interior looks blank—as if ice-covered even then. Modern cores from the Ross Sea floor, plus seismic-derived bed maps, make it **plausible** that riverine sediment once debouched there before modern conditions set in; the modern **Ross Ice Shelf** sits over basins that, ice-free, would be water.

2. **Buache (18th c.)** posits a **central frozen sea** and **through-continent waterway**, a design that—without being exact—**rhymes** with present-day bed reconstructions: depressions and troughs connect the Weddell, Ross, and Amundsen/Bellingshausen sectors. It's not a 1:1 match (his hypothesis was a hypothesis), but the concept of a **divided landmass** or archipelago emerges naturally from modern data.

3. **Modern subglacial hydrology** now confirms not only lakes but **river systems** beneath the ice, including a major basin-spanning river that drains toward the Weddell Sea—precisely the **kind** of process Buache used to justify an inland polar sea and that Finaeus's coastal "rivers" symbolically mirror.

These three strands don't prove that the early maps are copies of ancient polar surveys. But they **do** justify a careful, even generous reading: the maps encode geographic **ideas** that align—sometimes surprisingly—with what we now observe under the ice.

The Ross Sea Puzzle: A Case Study You Can Walk Through

Step 1 — Gather the old claim. On Finaeus's map, the Ross Sea sector shows **broad inlets** with **river-like** features. The suggestion: when the source map(s) were made, the coasts were **ice-free**, and a hinterland **drained** to the sea.

Step 2 — Ask what modern data says. Radar and BedMachine bed maps reveal that large swaths of West Antarctica's bed lie **below sea level**. The "western shore" of the Ross Sea is not a standing rock coast—it is a **depressed basin** that, if deglaciated, would be an ocean. In other words, "estuaries" here would make sense *if* the ice retreated.

Step 3 — Look for independent signals. Marine cores from the Ross Sea include layers of **well-sorted, fine sediments**, characteristic of **riverine** input from ice-free land. These layers end as **glacial** sediments dominate—indicating a transition from warmer, fluvial conditions to sustained glaciation. This supports the idea that **rivers once reached the Ross Sea**, even if the date, duration, and extent remain debated.

Step 4 — Decide what follows. The simplest synthesis is not that Finaeus drew satellite-accurate coastlines, but that his sources (however indirect) reflected **coastal sectors** of an Antarctica in some stage of reduced ice, feeding rivers to seas that occupied bedrock basins we can now map beneath the present ice.

What About Buache's "Split Continent"?

Modern bed maps confirm that East and West Antarctica are divided by the **Transantarctic Mountains** and that West Antarctica in particular is riddled with **below-sea-level** troughs. If sea level were held constant and the ice removed, you would not see one uninterrupted rock continent: you'd get **islands, embayments, and seaways**. Buache's inland sea and "cut" through the pole weren't cartographic prophecy, but his **model** of a **hydro-morphological** south—the idea that water basins and waterways organized the interior—**anticipated** the kind of structure modern scientists now map.

Historians caution that Buache's inland sea was a **theory**, not a transmitted memory. Fair. Yet to evaluate the deeper question—whether any "ancient knowledge" lurks behind such maps—we have to admit that the **broad shape** of Antarctica's hidden **physiography** ended up closer to Buache's **storyline** than to a solid monolith. That's not proof; it is a **data-driven invitation** to continue checking.

The Most Controversial Leap—and How to Avoid It

From here, some writers leap: if Finaeus and Buache align with hidden Antarctica, then ancient global mariners must have mapped the ice-free south thousands of years ago. That leap demands answers to a cascade of hard questions—**who** sailed, **when**, with **what** instruments, and how such precise coastlines were preserved. A wiser path is to **bracket** what the maps truly **show** and **don't**:

- They **do** show, at minimum, an unusual **coherence** in drawing coastal Antarctica with river mouths and inlets that make

Method Note: How to Compare Old Maps to Modern Data Without Fooling Yourself

- *Reproject the historical map to a modern grid and adjust for known projection quirks (e.g., double-cordiform).*
- *Compare relative positions and shapes, not absolute lat/long, because inherited source errors and scale inflation are common.*
- *Focus on diagnostic patterns (drainage logic; consistent estuary placement) more than precise coast placements.*
- *Weigh "wins" and "misses" equally; don't cherry-pick.*

hydro-geomorphic sense. They also embed ideas about an Antarctic **interior** that is water-organized, not uniformly rock.

- They **don't** give us consistent absolute positions, scales, or shorelines that survive pixel-to-pixel overlays; much is distorted by projection choices and compilation errors, not to mention the strong possibility of **conceptual** rather than empirical inputs—especially for Buache.

- Most importantly, **modern** discoveries—subglacial lakes, rivers, and bed topography—now show that certain **kinds** of features depicted or inferred on those maps **exist**; they just lie **beneath** a massive ice sheet.

Case Study: A Hands-On Comparison With Modern Data

Let's perform a simple, reproducible "desk experiment" to compare forbidden maps with satellite-era products.

1. **Choose the sector.** Start with the **Weddell–Ross transect** (west–east across Antarctica). On Finaeus, identify major **inlets** opposite each other on that sweep and any **river mouths** draining into them.

2. **Gather modern layers.** Download the public **BedMachine Antarctica** shaded relief and ice/land mask, and the location of the 2022 **subglacial river** routing toward the Weddell Sea (even a published schematic works).

3. **Normalize projections.** Reproject Finaeus's Antarctic panel to a **polar stereographic** grid and scale until one well-defined coastal "segment" (say, a concave bay and adjacent cape) best aligns with a modern counterpart. Don't force a global fit— **local** fits are what you're testing.

4. **Annotate patterns.** Mark a) paleo-estuaries on Finaeus; b) below-sea-level **basins** on BedMachine; c) the schematized path of the Weddell-bound river. You're not expecting

identity, only **structural** congruence: do inlets align with basins? Do "rivers" point toward pathways a real under-ice river could plausibly exploit?

5. **Score ruthlessly.** Note mismatches with the same enthusiasm as matches. The value here is **signal detection**, not romanticism.

Spatial Scoring Workflow

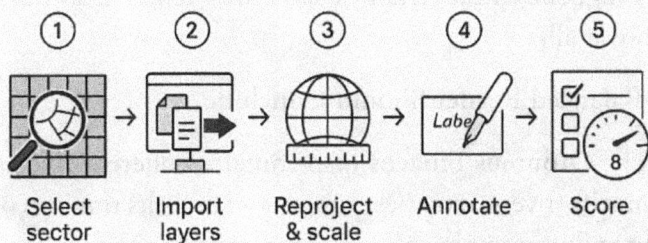

(1)	(2)	(3)	(4)	(5)
Select sector	Import layers	Reproject & scale	Annotate	Score

Spatial Scoring Workflow
(Five-Step Plate)

Where Did the Rivers Come From?

The new subglacial Antarctica invites a vexing but thrilling question: **when** did river systems last deliver sediment to polar seas **at scale**? Evidence from Ross Sea sediment cores indicates **fluvial** layers—fine, well-sorted silts and sands—before a switch to glacial deposits. That doesn't date to a mythical ice-free Holocene Antarctica; most geologists place ice-free **continental** conditions far earlier. But even within long glacial epochs, **warming pulses** and **grounding-line retreats** can create windows where rivers reoccupy coastal lowlands. The key is not to force the data into an all-or-nothing frame. A **mosaic** Antarctica—ice-free littorals, ice-gripped uplands—can reconcile fluvial signatures with a mostly glaciated south.

East vs West: The Hidden Archipelago

A final modern layer complicates and enriches the old maps: the **East/West** asymmetry. East Antarctica, high and cold, houses giants like Vostok; West Antarctica, lower and marine-based, sits in a **bathtub** below sea level for vast areas. If the ice vanished, West Antarctica would become a **sounding archipelago.** That's why a Buache-style inland sea makes **geophysically** more sense in the **west** than in the east—and why Finaeus's **vagueness** near certain "coasts" may reflect areas with **no true rock shore** at all.

What a Balanced Reader Should Conclude

- The **Oronteus Finaeus** map contains **coherent hydro-coastal motifs** (rivers, estuaries) and mountain belts ringing coasts that broadly resemble the real geography, once you account for projection and compilation artifacts. It doesn't prove ancient Antarctic mariners; it **does** justify deeper comparative work.

- The **Buache** map was a **theory-map,** but one whose **organization** (inland sea; east–west division) converges with modern **bed** geographies now revealed by radar and mass-conservation modeling. The specifics differ; the **structural** rhyme is noteworthy.

- **Modern discoveries**—from hundreds of subglacial lakes to a **460-km** under-ice river system—show that Antarctica is **hydrologically alive.** The presence of organized subglacial drainage lends physical plausibility to antique depictions of river-like features at the coast, even if the timescales are not the same.

- A careful investigator resists **overreach.** The maps could contain **echoes** of older knowledge **or** be the product of geographic reasoning and mythic expectations that—not coincidentally—mirror the **truth** of a water-organized Antarctica we only now can see.

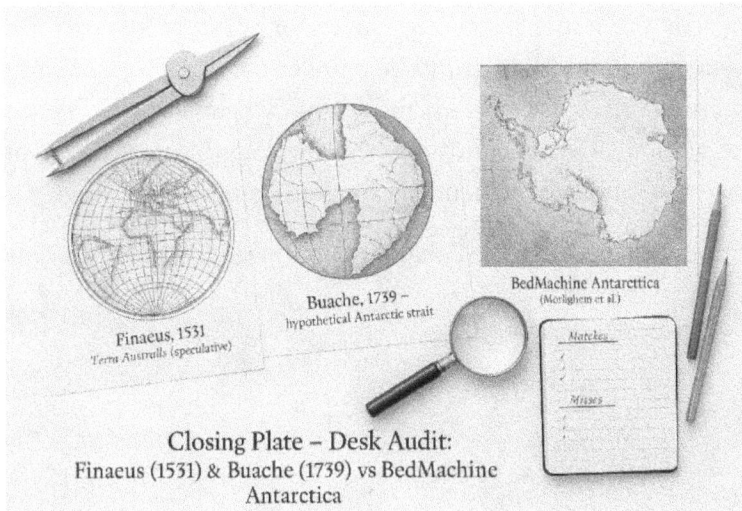

Finaeus, 1531
Terra Australis (speculative)

Buache, 1739 –
hypothetical Antarctic strait

BedMachine Antarctica
(Morlighem et al.)

Matches

Misses

Closing Plate – Desk Audit:
Finaeus (1531) & Buache (1739) vs BedMachine
Antarctica

Reading the Bed

- *Below-sea-level beds (blue on modern maps) imply future bays and channels in any deglaciated scenario.*
- *Sills and ridges matter: they would segment an inland sea into sounds and fjords—not one big bowl.*
- *Expect archipelago logic, not a single ring-shoreline.*

There's a reason cartography gets under your skin. A good map lets you move through **space**. A rare map lets you move through **time**. The Oronteus Finaeus and Buache maps have earned both defenders and skeptics because they invite a forbidden feeling: that somewhere in our tangled past, someone saw a version of the world that we, with all our satellites and supercomputers, are only now rediscovering.

Maybe that "someone" was a chain of compilers, each adding a scrap and a guess. Maybe it was an older maritime tradition whose coastal memory flickered into Renaissance hands. Maybe it was neither—just the lucky overlap between a **reasoned hypothesis** and the **messy truth** of a water-organized polar continent.

Whichever way you lean, here's what matters for our story: the **pattern**. Rivers in the ink of old maps meet rivers under the ice of new science. The gap between them is where curiosity lives, where you and I get to investigate without fear or agenda—where we keep the atmosphere of mystery and the muscle of evidence in the same room.

And that, more than any single "aha," is how forbidden maps become living maps.

Chapter 3

The Ancient Portolans

What makes portolans so intoxicating—so "forbidden," in the spirit of this book—is not just their beauty. It's the suspicion that they encode navigational know-how that appears to leapfrog the era that supposedly birthed them. Even their earliest surviving examples (late 13th–early 14th century) stand up uncannily well against maps produced centuries later. Unlike the theological mappae mundi or the tidy mathematics of the Ptolemaic atlases, portolans are practical, sea-worn, and—by many measurements—astonishingly accurate across the very regions where accuracy mattered most to sailors.

What exactly is a portolan?

At heart, a portolan is a mariner's working chart of coasts and harbors, typically focused on the Mediterranean and Black Sea (later expanding outward with world "portolans"). Their signature features include:

- Dense networks of straight "rhumb lines," radiating from compass roses (often 16 or 32 winds), were used to steer consistent bearings.

- Coastline detail optimized for pilotage: headlands exaggerated, shoals and reefs marked, river mouths emphasized, place names written perpendicular to the shore so they read from the sea.

- Execution on vellum with color conventions (e.g., major toponyms in red, lesser in black; wind-line palettes that help the eye parse the wind system).

All of this turns a flat animal skin into an instrument. It doesn't impose a classical latitude-longitude grid, yet when measured—especially across the inland seas—many portolans achieve surprisingly tight relative positions. A celebrated example: the 1339 Dulcert chart puts the total longitude of the Mediterranean-plus-Black Sea within roughly half a degree of modern values, a figure good enough to make any glass-and-brass 18th-century navigator proud.

Were portolans "centuries ahead of their time"?

That phrase can mislead—yet it captures a genuine puzzle. From a "tools" standpoint, the Age of Discovery was still centuries away from the marine chronometer that made longitude routine. So how did 13th- and 14th-century chartmakers place coastal features with such consistent fidelity—especially east-west—across the Mediterranean?

Mainstream explanations emphasize cumulative pilotage: generations of coasting distances and compass bearings, culled from pilot books (portolani), refined in practice, and codified by superb draftsmen. The widespread European adoption of the magnetic compass by the late 1200s is often credited with enabling the rhumb-line system seen on these charts (and the rhumb network certainly helps a crew hold a course).

Yet some aspects still nag. Analytical reconstructions suggest that, in some cases, the rhumb frameworks on influential charts are not merely decorative or heuristic, but mathematically consistent. In several studies of early modern portolan-style maps (including the famous 1513 chart compiled in Ottoman waters), the wind-line networks and

Portolan Chart, Working Definition: A medieval or Renaissance nautical chart drawn on vellum, optimized for real-world navigation with rhumb-line wind networks, true-to-practice coastline detail, and legible harbor labeling—prioritizing what a pilot needs, not what a philosopher prefers.

their centers behave as if derived from an underlying geometric scheme: when transformed from a circular (polar) construction into a rectangular grid, many prominent points fall plausibly where modern lat-long would place them. That implies, at minimum, a sophisticated geometric backbone beneath the practical surface, and at maximum, an inheritance from older, more technical source charts.

The Mediterranean web: a network that breathes trade

Portolans were as much instruments of commerce as they were of navigation. Each rose, each rhumb, is a choice—a fishing ground here, a dangerous shoal there, a day's sail between trusted ports. As the network expanded beyond the Pillars of Hercules, the charts began to hint at something bigger: a sea-road culture stretching into the Atlantic, down the West African bulge, and, eventually, around Africa into the Indian Ocean.

The mainstream picture is breathtaking enough: by the early 1500s, Iberian powers stitched together secret atlases and guarded rutters, creating a proprietary knowledge economy from Brazil to Malabar. But certain maps, produced earlier than we would expect them to display such detail, tempt a more radical reading: that there existed, somewhere in the archive of coastal lore and pilots' scribbles, base maps of unusual scope—sources the Renaissance cartographers could dip into long before their national fleets had finished exploring those waters.

Let's test that idea through two case studies.

The Portolan Paradox: No classical graticule—yet world-class accuracy where it counts. Are portolans simply the masterpiece culmination of pilotage, or do they incorporate fragments of an older, more mathematical cartography?

Case Study 1: The Cantino Planisphere — a smuggled world

On 19 November 1502, a magnificent world chart reached Italy under clandestine circumstances. Its path was furtive by necessity: the state that nurtured it punished map-smuggling with death. The chart in question—the planisphere associated with Alberto Cantino—displays, in rich pigment and gilded iconography, a partitioned Atlantic (the Treaty of Tordesillas line features prominently), a confident sweep of West Africa, the bulge of Brazil newly claimed—and, puzzlingly, a nuanced outline of parts of India that Portugal, by official reckoning, had not yet charted with such care.

The timing matters. Crafting a chart like Cantino's likely took many months. Internal heraldry pins its knowledge to events of 1500–1501 (Brazilian landfall; specific flag icons over Cochin and Cananor on India's southwest coast), yet it portrays northwestern and eastern India with a competence that outpaces expeditionary records. For example, Sri Lanka is placed with a closeness in size and position that anticipates later sailing visits; the eastern littoral of India is shaped far better than the clumsy Ptolemaic tradition could account for. How? The ships thought to have supplied this intelligence—the first Portuguese fleets

Zeno in Two Columns

Mainstream View: A 1558 publication with dubious pedigree, probably derivative, possibly patriotic fiction; the "ice-free Greenland" motif is not evidence of prehistoric surveys.

Alternative Reading: Elements of the sheet align with rigorous frameworks (polar and portolan grids) and encode realistic coastal relationships—hinting at an older, technically capable source tradition that the 1558 editor poorly re-projected.

to round Cape Comorin and prowl the Bay of Bengal—don't show up in the logs until 1505 and after.

There are orthodox escapes: pilots from the Red Sea or Gujarat might have traded charts; Arab and Indian seafarers possessed deep, empirically tested knowledge; fragments of earlier, non-European charts could have circulated in coastal marts from East Africa to Kerala. All are plausible—and indeed likely contributors. Yet the Cantino chart still reads like an anthology: European symbology and claims pinned over baselines that, in places, feel older than the voyages that supposedly generated them.

Case Study 2: The Zeno Map — evidence or invention?

A two-century riddle hangs over a map published in Venice in 1558. Its editor claimed it originated in 1380, drawn from the travels of two Venetian brothers who voyaged far into the North Atlantic. Modern historians often file it under "pseudo-history," or at least "unreliable hybrid," suspecting that the 16th-century editor conflated hearsay, later charts, and patriotic embroidery. That skepticism is reasonable and widely held.

And yet—set aside the editorial backstory and look at the thing with a navigator's eye. The chart shows Iceland and Greenland, but also detailed continental margins from Norway to Scotland, the Faroes, and the Shetlands included. Technical analyses have found that, when

Cantino's Anomaly in One Breath: A 1502 world map stitches Portuguese banners to Indian coasts with a precision the official timetable doesn't quite earn, implying that cartographic intelligence arrived from beyond the documented voyages—most likely via pre-existing regional charts and pilots, but perhaps in part via older (now-lost) compilations.

reconstructed on a consistent polar framework, many latitudes and longitudes across the sheet behave as if the chartmaker knew how far east-west the North Atlantic runs at given parallels—and how north convergences should curve. In one reconstruction, the map becomes coherent on a polar grid; in another, it fits a portolan-type "two-norths" grid (a square rhumb system rotated against a spherical baseline), suggesting a mathematical substrate beneath the 16th-century editor's re-projection.

More provocatively, Greenland is drawn as if ice-free, with interior mountains and river systems reaching the sea. That detail has been interpreted by some analysts as a faint memory of deep antiquity—of surveys (or source maps) made in an epoch before the modern ice cap took its current form. That's an extraordinary claim, and it needs extraordinary caution: creative medieval mapmaking can invent mountains; post-glacial isostatic changes can alter coastlines; a later editor could have imposed a fanciful style. But the Zeno map keeps surfacing in the literature because when you test it—mathematically rather than stylistically—portions resolve into patterns not easy to dismiss.

Navigational maps "centuries ahead": what that really means

The phrase "centuries ahead" is rhetorical. What we truly mean is that portolans and portolan-style world charts often display: (1) an empirical mastery of Mediterranean pilotage unrivaled in their age; (2) a geometric discipline that, in places, hints at mathematical thinking beneath the ink; and (3) a willingness to compile from any source— pilots, merchants, rival states, and, perhaps, older chart traditions whose originals are now lost.

A crystallizing example lives in the Ottoman-world chart compiled in 1513. Its annotations famously mention multiple source maps, some attributed to classical times, and include the remark that several were "based on mathematics." Whether we take every annotation literally or not, the map behaves in technical analyses as though its rhumb system is anchored to a specific mathematical center (plausibly where a major meridian and tropical line cross), and can be translated into a modern grid with coherent results. That's exactly the sort of outcome you'd expect if a compiler were layering practical pilotage onto a formal geometric framework inherited from earlier, more abstract cartography.

"Mediterranean and global trade routes that shouldn't exist"

At first glance, this sounds like provocation. But the phrase captures a set of anomalies—places where the map record appears to be slightly ahead of the voyage record, or where a European map shows regional detail without a clear European reconnaissance behind it.

The most reasonable explanation is *not* time travel; it is contact: knowledge spills across pilot communities. A Gujarati pilot doesn't need a European expedition to know where the river mouths sit on the Coromandel Coast; a Yemeni captain from Aden doesn't need a Lisbon imprimatur to sketch the shoals of Socotra. When a European collector assembles these fragments into a world sheet, the result can look like clairvoyance.

But there's a second layer. In the literature on portolan-style maps, several analysts have argued for underlying "projections before projections"—cartographic grammars that predate the modern graticule. If fragments of those grammars found their way into Renaissance compilations (via libraries, copied atlases, or pilot charts transmitted through Mediterranean brokerage hubs), some "routes that shouldn't exist" are simply routes someone mapped long ago whose parchment didn't survive. On the margins of Antarctica, for example, a cluster of 16th-century maps portrays a southern land with mountains and river mouths. Some readers interpret this as fanciful; others note intriguing overlaps with modern sub-ice topography and keep the question open.

How were portolans made?

Think of three strands woven together:

1. **Itinerary knowledge:** Coasting distances timed in watches, bearings read by compass, seasonal winds memorized (the *tramontana*, the *sirocco*, the monsoon).

2. **Workshop geometry:** Grids of rhumbs laid with compasses and straightedge, repeated across the sheet from multiple rose-centers, producing a stable lattice pilots can use underway.

3. **Compilation culture:** Harbor masters, merchants, privateers, and pilots all contribute fragments. A chartmaker is a broker of small truths.

The result is not a "map of the world" in the modern sense. It's a field instrument that grew fat with the world as the outer seas were explored. By the early 1500s, some of these instruments, like Cantino's, had become propaganda as well—declaring claims, projecting power, and [quietly] showing more than the official record could justify.

The deep question behind the charts

Every good map is a bet on reality. Portolans bet, again and again, that the world would repay close observation, careful compilation, and practical geometry. They were right. But hidden in their success is a whisper: *Where did the geometric discipline come from?* Was it born in the shipyards and benches of Italian ports and Majorcan workshops alone? Or did it rework older schemata—perhaps preserved in monastic libraries, Alexandrian remnants, or Islamic and Byzantine compilations whose own antecedents reached back further still?

This book, *Forbidden Maps,* is not here to push dogma. It is here to hold the puzzle steady while we turn it in the light. A coherent way to think about portolans—one that respects both mainstream caution and alternative curiosity—looks like this:

- **Baseline:** Portolans are the pinnacle of practical medieval cartography, generated by cumulative pilotage and refined workshop methods.

- **Overlay:** Some world-scale portolan-style charts incorporate geometries and regional accuracies that imply access to prior cartographic grammars and non-European chart traditions.

- **Edge cases:** A few anomalous features—like unexpectedly good outlines far from Europe, or southern landmasses with structured interior features—invite testable hypotheses about lost sources. In several notable analyses, conversions of rhumb networks to modern frameworks yield structured correspondence, which deserves formal study rather than preemptive dismissal.

In the lamplight, the pilot folds the vellum and thumbs the crease where he's drawn a tiny dot for a harbor—one he's never visited, but whose soundings he trusts because someone, decades ago, wrote them down in the same measured hand as the rest. That is the spirit of the portolan: trust the machine. It is, in that sense, the opposite of mystery. And yet, when we pull back, mystery regathers at the edges: charts that know too much, too soon; geometries that convert too neatly to be merely decorative; southern coasts that mirror ice-hidden outlines.

The honest position is not credulity or orthodoxy. It's curiosity with standards. As we move deeper into this book's thesis—that old charts sometimes preserve fragments of older worlds—we'll keep those standards in view. The Ancient Portolans do not need to be "from Atlantis" to shock us; they only need to be as good as they are. And they are very, very good.

Part II: The Power Lines of the Earth
Chapter 4: The Mystery of Ley Lines

You've seen it yourself on a good map: draw a ruler-straight line and you can hit a surprising number of ancient places—stone circles, burial mounds, hillforts, old churches, standing crosses, even prehistoric trackways. That simple, unsettling observation is the heart of the "ley line" idea. This chapter treats it seriously but not uncritically. We'll look at the best cases for ancient straight-line planning, the most common alternative explanations, and what modern tests actually show. We'll range from Stonehenge to the Great Pyramid, from Inca "line systems" to the Peruvian desert, and then dive into a focused case study on the person who put "leys" on the map: Alfred Watkins.

What are "ley lines," exactly?

In the narrow, historical sense, "leys" were Alfred Watkins's claim that many ancient features in the British landscape fell on deliberate, ruler-straight alignments—basically old sighted trackways used for travel, trade, and surveying. He never argued for "mystical energy" running along them; that came later with the 1960s–80s Earth Mysteries revival. Watkins's method was plain: use large-scale Ordnance Survey maps, mark ancient points, and look for long, straight runs of three or more points. He then checked them on foot to see if evidence of a track or sighting markers survived.

In the broader, modern sense, people use "ley lines" to mean just about any straight-line connection between ancient or sacred sites. That broader usage bundles together very different things:

- **Sighting lines** and **trackways** (the original Watkins idea).

- **Processional/sacred lines** like the Inca **ceques** radiating from Cusco which were administrative and ritual lines organizing shrines across the landscape.

- **Astronomical alignments** (e.g., Stonehenge's solstitial axis; lunar or solar lines marked by avenues).

- **Geodetic/"world-grid" ideas** linking far-flung monuments by longitudes/latitudes or geometry (a controversial, big-canvas claim we'll parse later).

- **"Earth energy" lines** and dowsing claims (the Earth Mysteries strand, subject to several instrumented studies, with mixed outcomes).

This chapter keeps those categories distinct so we can assess each on its own merits.

Why straight lines at all?

Straight lines solve problems. If you need to cross country quickly and repeatedly, straight is efficient. If you need to aim at a distant hilltop, a sunrise notch, or the exact azimuth of midsummer sunrise, straight is essential. If you need to lay out land boundaries, re-survey fields after floods, or triangulate locations, straight lines and consistent measures are your friends.

Ancient planners clearly did all three: travel, ritual, and measurement. In Peru, the Nazca straight lines run for miles—some bisect older animal figures, suggesting phases of use, with plausible ritual functions tied to processions, offerings, and sky-watching. At Cusco, the **ceque** network assigned shrines along lines to specific kin groups—an administrative and ceremonial grid across the landscape. In Western Europe, megalithic builders standardized modules (e.g., the "megalithic yard") and used careful astronomy and geometry in site planning.

Even ancient geodesy shows up. When French surveyors mapped Egypt after 1798, they noticed the Great Pyramid is exquisitely aligned to the cardinal directions and sits as a superb geodetic marker; they used its meridian as a mapping baseline and noted how its diagonals encapsulate the Nile Delta. Mainstream surveys confirm the Great Pyramid's orientation is within just a few arcminutes of true north.

Theories of Ancient Alignments Linking Sacred Sites

The pragmatic trackway theory (Watkins's core idea)

Watkins argued that many British "ancient things"—mounds, moats, standing stones, beacon hills, pre-Norman churches—often line up straight because people laid out sighted routes that threaded known landmarks. He published *Early British Trackways* (1922) and then *The Old Straight Track* (1925), providing mapped examples and field observations. He drew on earlier work, noting astronomical alignments but cast leys primarily as usable lines of sight and movement.

How that holds up:

- Britain *does* preserve traces of straight prehistoric and Roman roads.

- On the ground, some mapped leys still coincide with lanes, parish boundaries, or ridge lines; others don't.

Why straight matters

- *Straight lines are practical (fast travel, unambiguous re-survey).*
- *Straight lines are ritual (processions, sacred routes, shrine networks).*
- *Straight lines are astronomical (precise sunrise/sunset or star bearings).*
- *Straight lines are geodetic (triangulation, mapping baselines).*

- Statistical work shows that when you scatter many points (sites) on a map—especially in a landscape saturated with churches and barrows—some long straight alignments arise **by chance.** That doesn't mean all alignments are chance, but it raises the bar of proof; you must show a line was intended and used, not just "can be drawn."

Sacred networks that happen to be straight

The Inca **ceque** system is a crisp example: invisible "lines" radiated from the Temple of the Sun; along each line, specific shrines (wak'as) were maintained by designated groups. That's a straight-line organizing principle, but it's **ritual-administrative**, not about magical beams.

In Scotland's Callanish complex, straight stone "avenues" lead to a central ring; other alignments point to lunar or solar extremes across the landscape. The venue is straight, the meaning is astronomical and ceremonial.

In Malta's Mnajdra complex, sunlight penetrates to specific interior targets on solstices and equinoxes with remarkable precision; you can literally "watch" a light-blade sweep between set stones across the year. That's a moving alignment scaffolded by architecture.

Geodetic and "world-grid" proposals

Some researchers plot long-distance relationships between major sites using longitudes/latitudes and base-3 or precessional number sequences. Examples include proposed relationships among Giza, Angkor, Tiruvannamalai, and a megalithic site in Taiwan—arranged at intervals of 24°, 48°, and 90° east of Giza, numbers that also appear in ancient mythic numerology and precessional cycles. These ideas are boldly synthetic and invite skepticism; still, the claims make testable geometric statements about spacing and bearings.

Other geodetic claims spotlight Egypt itself: the Great Pyramid's cardinal accuracy is undisputed; more speculative are the ideas that it marks special global "centers" or that its placement was intended to serve as a master survey point for the Delta's extent. Those arguments come from cartographic and historical readings and deserve to be labeled **hypotheses**, not facts.

"Earth energy" along lines

From the 1970s onward, some investigators proposed that lines weren't just sight lines but **energy** conduits. The Dragon Project at the Rollright Stones tried to instrument stone circles—looking for anomalous magnetism, radioactivity, ultrasound, etc. Reports and retrospectives stress that the best-controlled measurements didn't find clear, repeatable "new forces," though localized geological effects and mundane magnetism/radioactivity can vary. In short, interesting anomalies sometimes appear; a consistent energy-line mechanism remains unproven.

From Stonehenge to the Great Pyramid: Coincidence or Design?

Stonehenge

Nobody seriously doubts that Stonehenge encodes solar alignments. On the summer solstice, sunrise appears in line with the monument's axis near the Heel Stone; on the winter solstice, the sunset aligns on the opposite bearing. That's straight-line planning with precise seasonal targets, not a "map of the world" or an energy grid.

The Great Pyramid

Cardinal accuracy of the Great Pyramid is extraordinary (deviation just a few arcminutes from true north–south). There are viable methods to achieve that—stellar transits, equal shadow techniques near equinox, etc.—and multiple technical papers discuss them. Modern geodetic surveys confirm the precision; what's debated is the *purpose* of that precision.

Historical mapping anecdotes add to the mystique: French savants used the pyramid as a mapping reference; on certain reconstructions, its diagonals frame the Nile Delta neatly in plan. Those reconstructions imply a conceptual geometry for Egypt as a 7-degree "strip," with the pyramid marking key reference meridians. This is evocative and has been argued in detail, but should be presented as an interpretive model, not an agreed-upon fact.

Global alignments: how far can we push?

Claims that Giza, Angkor, and other sites fall on elegantly spaced longitudes or share precessional number spacing make for an exciting hypothesis about long-baseline surveying. The sober way to handle them: (1) state the math precisely, (2) check with modern coordinates and great-circle bearings, (3) run chance-expectation comparisons on many random site sets. Some published measurements do land

How to judge a proposed "ley"
- *Intentionality: Is there independent evidence the line was meant (paths, sighting stones, way-marks, historical notes)?*
- *Function: Travel? Ritual? Astronomy? Administration? Mapping?*
- *Visibility: Are inter-points intervisible or plausibly signaled?*
- *Statistics: Is the line more impressive than chance given site density? Use a pre-registered test.*

impressively close to exact integers of degrees, but given a global search space, chance alignments do appear. Treat these as **research prompts**, not settled history.

Nazca: straight lines, sacred traffic

At Nazca, very long, very straight geoglyphs coexist with older animal figures. Straight lines intersect the landscape relentlessly, and studies argue for processional use, shrine-by-shrine offerings, and astronomical references (e.g., potential Orion connections in specific figures). The desert's physics—pebble varnish, gypsum, and vanishingly low rainfall—helped the straightness survive. None of this requires "airstrips" or spacecraft. It does require straight-line intent.

Case Study: Alfred Watkins and *The Old Straight Track*

The "revelation." In 1921, near Blackwardine (Herefordshire), Watkins later said he suddenly "saw" the landscape as threaded by straight lines that linked ancient markers. He went home and drew lines across his maps; more and more alignments seemed to appear. He then wrote two books: *Early British Trackways* (1922) and the more systematic *The Old Straight Track* (1925). The latter catalogs alignments, field methods, and examples, arguing that "leys" were old sighted routes connecting "mark stones," mounds, moats, beacon hills, and early churches.

What Watkins did well

He insisted on **field-checking** with his camera and boots. He cared about **use**—tracks that actually went somewhere. He looked for **continuity** (prehistoric sites later "caught" by churches, crosses, and parish boundaries) and for **function** (beacons intervisible on ridge lines; lanes aligned to churchyard crosses). That's disciplined landscape archaeology for his time.

Where it drew fire

Archaeologists complained that post-Roman and medieval features got mixed with prehistoric ones; that cherry-picking points among thousands guarantees some lines; and that most "leys" lack independent evidence of intentional straight route planning. Statisticians later showed that when you randomly sprinkle points, long straight runs happen surprisingly often—so you must pre-register your test and dataset to beat the "power of chance."

The 1960s–80s revival

Watkins's practical ley morphed in popular imagination into "energy lines." Writers and dowsers proposed that standing stones tap currents, that circles act like acupuncture needles, and that leys carry a subtle flow. The Dragon Project tried to measure such effects at the Rollright Stones; after years of monitoring, the safest summary is that geology matters (some stones are mildly radioactive or magnetic), but a stable, exotic "earth energy" detectable as a line remains unproven.

Key facts: Stonehenge & Giza
- *Stonehenge's axis frames summer-solstice sunrise/winter-solstice sunset. That's intentional design.*
- *The Great Pyramid is aligned to the cardinal directions within a few arcminutes. Precision: yes. Energy beams: no evidence.*

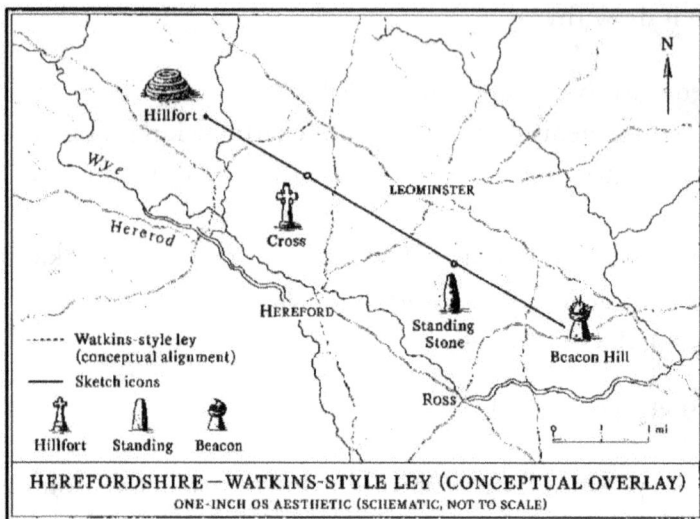

HEREFORDSHIRE — WATKINS-STYLE LEY (CONCEPTUAL OVERLAY)
ONE-INCH OS AESTHETIC (SCHEMATIC, NOT TO SCALE)

How to Test a Claimed Ley Like an Adult (A Field-Ready Guide)

1. **Define the hypothesis before you look.**

   o Pick a **fixed region** (e.g., one county).

   o Pre-list **feature classes** allowed (e.g., Neolithic long barrows, Bronze Age round barrows, Iron Age hillforts, early churches with known pre-Christian origins, standing stones).

   o Define **tolerance** (e.g., within 100 m of the line for rural features; tighter in dense parishes).

   o Set a **minimum length** (e.g., ≥10 km) and **minimum hits** (e.g., ≥5 qualifying sites). Why: This lets statistics mean something.

2. **Run a chance baseline.**

o   Monte Carlo: randomly rotate/translate the same line across the same map **1000 times**; record how often ≥5 hits occur. If your observed alignment is rarer than, say, 1% of the random runs, that's interesting.

3. **Check intervisibility and practicability.**

    o   Use a modern digital elevation model (DEM) to see if the sites could be seen in sequence.

    o   Walk or drone-scan key spans: is there an ancient terrace, hollow-way, parish boundary, or bank in the right place?

4. **Look for *independent* function.**

    o   Travel: Is there soil compaction, metalling, or historic way-markers?

    o   Ritual: Does the line match solar/lunar azimuths on key dates?

    o   Administration: Does it slot into a ceque-like assignment or boundary system?

Watkins in one minute
- *1921: Sees straight landscape alignments near Blackwardine.*
- *1922: Early British Trackways; 1925: The Old Straight Track.*
- *Method: map first, walk second; focus on useful lines.*
- *Legacy: sparked a century of debate—from sober trackways to speculative "energies."*

5. Instrument, but keep it sober.

   o If you want to test "energies," log background magnetism, radioactivity, infrasound, and ultrasound along the line *and* along nearby control transects. Expect geology—not magic—to dominate.

Comparative Alignments: Beyond Britain

Inca Ceques (Cusco)

The ceque system is the most explicit ancient "line network" we have: radiating from a center, organizing hundreds of shrines in straight lines that structured ritual calendars and social duties. This isn't "secret energy" talk—it's documented sacred geography and administration.

Callanish and other megalithic complexes

Callanish combines a central ring with straight "avenues" and multiple astronomical targets (moon, sun, possibly notable star risings), showing that linear planning could scaffold repeatable sky events.

Egypt's geodesy

Accounts tying the Great Pyramid into a geometric conception of Egypt—meridians, delta triangles, and long-baseline references—are elegant. They blend hard facts (cardinal accuracy) with interpretive claims (national geometry). Keep the boundary clear: **measured orientation** is certain; **grand-design geodesy** is an attractive hypothesis that needs independent archival or field proof.

Global mapping anomalies and ancient cartography

Separate from "ley" talk, a body of work on early maps argues that some medieval/Renaissance charts encode far older source knowledge, including accurate longitudes and sophisticated projections—well before the modern chronometer era. If even partly true, that implies a deep tradition of measurement and mapping—precisely the skills behind long, straight alignments. Consider it a context, not a conclusion.

Skeptic's Corner: Chance, Cherry-Picking, and Clean Methods

The biggest pitfall in ley hunting is **retrofitting**: you notice a couple of sites, then "find" others along the way and let the line drift to catch them. That turns statistics into mush. The right way is to **lock** the line first (bearing, start/end), then test it against a fixed, relevant site list. S. R. Broadbent's work modeled the expected number of k-point collinearities in random point fields; it shows why dense landscapes can yield impressive-looking "leys" by accident. The remedy is not despair—it's a method.

Another pitfall: **mixing time periods.** A Bronze Age barrow, a Saxon church, and an 18th-century folly might all be in a straight line. That can still be historically meaningful (later builders often repurpose older sacred places). But if you're claiming prehistoric route planning, your **core evidence** must be prehistoric.

> ### Pre-registration checklist
>
> *Region fixed? Feature classes fixed? Tolerance set? Minimum length/hits set? Randomization plan written? If yes, you're doing it right.*

Finally, the **energy** question. The most thorough thing one can say is that careful projects have not produced a robust, reproducible "new field" distinct from known geology/geophysics along straight-line corridors. Interest remains; proof doesn't. That's okay. Curiosity + rigor beats certainty without data.

Practical Walkthrough: Re-creating a Watkins-Style Survey (Herefordshire Example)

1. **Map session (two hours).**

Open a 1:25,000 or 1:50,000 map for a 20×20 km square around Blackwardine. Mark: scheduled barrows/henge sites; hilltop churchyards with known early origins; standing stones/crosses; beacon hills. Draw only **one** trial line: choose a bearing that naturally links two high-status endpoints (e.g., a hillfort to a beacon). Extend it across the square and note hits within 100 m.

2. **Statistics (one hour).**

Use a simple script or an online tool to rotate/translate that line across your square 1000 times. Count how often ≥ the same number of hits occur. If your line is rarer than 1% of cases, flag it for fieldwork.

3. **Field day.**

Walk at least two inter-point spans. Look for hollow-ways, aggers (road embankments), lynchets, boundary banks, parish stones, or a churchyard cross aligned to the approach. Photograph sightlines and take horizon bearings.

4. **Astronomy cross-check.**

Get azimuths for solstice sunrise/sunset and lunar extremes at your latitude. If the line matches within a degree or two and you have ritual features, you've got a candidate sacred alignment; if not, it may be a travel line.

5. **Optional instrumentation.**

Log background magnetism and gamma counts at 50-m intervals on the line and on a nearby control line. Expect geology, not a miracle; look for mundane correlations (ironstone, granite, faulting).

Three rules that save you
- *Fix your test before you hunt.*
- *Keep periods consistent for claims of origin.*
- *Use controls (nearby lines, randomized lines, blind instrumentation).*

1. Sightline: hillfort →

2. Boundary stone

bearing of alignment

3. Field lynchet

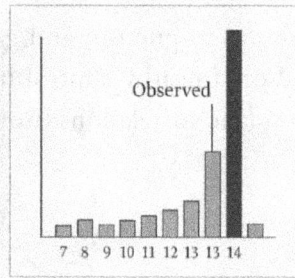

Observed

7 8 9 10 11 12 13 13 14

Example randomization test
(illustrative data)

What Ley Lines Are—and Aren't

- **They are** a powerful way to **notice** patterns of intentional alignment: some for travel, some for ritual calendars, some for sky-watching, some (perhaps) for surveying and land geometry.

- **They are not** a license to connect anything to anything. Without controls and period discipline, impressive lines appear by accident—especially on feature-dense maps.

- **They can be** tools for co-discovery. When you invite readers to test lines themselves—pre-registering the plan, walking the land, and publishing negatives as well as positives—you get closer to history and away from wishful thinking.

It's tempting to believe that one grand system explains every straight alignment on Earth. The data say otherwise. Ancient people used straight lines because straight lines **work**—for moving across rough country, for catching the sun precisely, for laying out shrines and boundaries, and for making maps that tie land to sky. That's already wondrous enough. The responsible move is to test lines the way Watkins would recognize: map first, walk second, and keep your standards high. When a line survives that gauntlet—when it's longer than chance expects, intervisible, practical, and functionally anchored—you can start calling it meaningful history rather than a line on paper.

Your shortest path to real findings
- *Pick one county.*
- *Pick one hypothesis (travel, ritual, or astronomy).*
- *Pick one line. Pre-register. Test. Walk. Report, even if the result is "not significant."*

Chapter 5

Global Grids and Earth Energies

Y ou don't have to believe in anything mystical to feel the pull of certain places. Stand on a granite tor at dawn, or inside a ring of weathered stones at dusk, and your body quietly reports: *something's different here.* The question is what. Is it geology? Electromagnetism? Ancient planning? Cultural storytelling that primes us to sense significance? Or do these places sit on a genuine, planet-spanning framework—a grid—shaped by energies our textbooks don't yet fully capture?

In this chapter, we'll walk you through the strongest ideas behind a planetary energy network, how megaliths and monuments appear to "snap" into wide-angle geometric patterns, and a hands-on case study of the Becker–Hagens "Earth Star" grid. We'll keep the tone grounded, highlight what's testable, and note where mainstream scholarship nods—or shakes its head.

The Theory of a Planetary Energy Network

At its core, the "Earth energies" idea claims that the planet expresses large-scale structure: lines, nodes, and lattices that influence where humans sense place-power and—more practically—where cultures choose to put stones, temples, and cities. Advocates say these structures are not random; critics point to coincidence and cartography tricks. Let's unpack the main pillars.

Ancient long-range knowledge and mapping hints

If prehistoric or early historic cultures knew more about global geography than we assume, you'd expect echoes of that knowledge to survive in maps, measures, or alignments. And in fact, certain medieval and Renaissance charts appear to carry data far in advance of their time, implying older sources behind them. For example, the famous Piri Reis chart (1513) sets Africa and South America in surprisingly correct relative longitudes—long before marine chronometers made longitude easy, which suggests underlying prototypes preserved through copying chains.

Broader dossiers discuss 14th- to 16th-century charts with modern-like precision in latitude/longitude and even debated portrayals of polar coasts. The thread that ties them together is the possibility of earlier cartographic knowledge—perhaps from a seafaring culture whose charts were recopied across centuries. Some researchers explicitly connect these controversies with Ice Age coastlines and lost shore cultures later drowned by rising seas.

2) "Earth energies" in physical terms

Strip away metaphors and you still find real geophysics that could create place-specific anomalies:

- **Telluric currents:** natural, low-frequency electrical currents flow through the ground and oceans, driven mostly by solar/geomagnetic interactions. They're standard tools in magnetotellurics, used to image the subsurface and fault zones. (Definitions and uses are mainstream geophysics.)

- **Piezoelectric rocks:** stress on quartz-rich rocks can generate electrical fields; laboratory and geological literature documents piezoelectric effects in quartz and mylonites, with recent work

Certain pre-modern maps contain latitude/longitude relationships too accurate for their era, implying far older sources behind the charts. This is a testable claim about geometry on paper, not belief.

even tying stress-electricity to gold deposition in quartz systems.

These phenomena don't "prove ley lines," but they do show that landscapes can host measurable electrical and electromagnetic variations. It's reasonable to ask whether some sacred sites coincide with spots where geology naturally focuses subtle fields.

3) Cultural patterning: lines, paths, and alignments

Long before modern talk of "leys," many cultures inscribed lines into landscapes for ritual mobility and memory. Yet in Britain and Europe, the 20th-century idea of straight, multi-mile "leys" linking ancient sites has been fiercely debated. Some archaeologists and statisticians argue that with enough points on a map, straight alignments will appear by chance; other researchers, including careful ley historians, shifted the conversation from "mystic rails of power" to more nuanced landscape phenomenology and sightlines.

The pragmatic takeaway: we should analyze alignments with robust methods (GIS, Monte Carlo tests) and separate three questions: (a) *Are alignments real beyond chance?* (b) *Were they intentional?* (c) *If intentional, why—astronomy, visibility, procession routes, or something energetic?*

Telluric currents and piezoelectric effects are real physics. The open question is whether ancient builders intentionally targeted places where these effects are unusually strong.

How Megaliths and Monuments Fit Into Geometric Patterns

Zoom out from a single stone circle, and patterns begin to suggest themselves—ovals, triangles, long axes, and inter-site roads. Several lines of inquiry matter here:

Field-scale geometry and "yardsticks"

Meticulous surveys of British megaliths led to proposals of consistent measures and astronomical orientations; later critiques challenged the statistics, yet even skeptical reviews concede non-random structure at some complexes. The fair reading today is that some builders did impose repeatable geometries and sky-aware planning, but a universal "megalithic yard" remains contested.

Landscape-scale systems

In northern France and southern Britain, researchers have argued for extremely broad, planned ovals and corridors linking multiple megalithic hubs—Stonehenge and Avebury among them—hinting at centralized planning that extended across regions. This view reframes megaliths as components of a civilization-wide project, not isolated monuments.

World-scale hints

When you lay down catalogues of "First Places"—the earliest urban centers, temple complexes, and persistent sacred landscapes—some investigators see symmetry and long chords that *might* follow great-circle arcs. Others say that's human pattern-seeking. What's not controversial is that ancient large-scale mapping or memory could survive in fragmentary ways—like map traditions that resurface centuries later. Several long sections in the Underwater-Cities/Ancient-Maps literature dive exactly into this continuity problem.

Case Study: The Becker–Hagens "Earth Star" Grid

The Becker–Hagens grid (often called the "Earth Star" or UVG-120) is the most widely referenced global grid model in Earth-mysteries circles. Here's what it is—and isn't.

What the model proposes

- A **polyhedral net** (icosahedron/dodecahedron transformations) inscribed on the globe, subdivided into **120 identical triangles** and ~**62 intersections** ("nodes"), forming a skeleton that purportedly relates to recurring "anomalous" zones, tectonics, ocean gyres, and clusters of ancient sites.

- The lineage of the model is credited by its authors to earlier "vile vortices" work that plotted 12 anomaly zones (e.g., Bermuda and its antipode) and to New Age–era syntheses that mixed geometry, cartography, and mythology.

- In their long essay "The Rings of Gaia," the authors connect the grid's **15 "rings"** to a mapping scaffold and argue that medieval charts (including Piri Reis) preserve hints of this geodesic pattern.

How it's built:

1. Start with an icosahedron/dodecahedron relationship on a sphere.

2. Subdivide into a **hexakis icosahedron** (your 120 triangles).

> *Pattern or pareidolia? The only way to know is to publish datasets and run blind statistical tests—great circles vs. randomized controls; a priori criteria vs. post-hoc cherry-picks.*

3. Georeference the network and inspect whether known anomalies, tectonic features, currents, or monumental clusters sit near nodes and lines.

Earth Star Grid (UVG-120),
Geometry-Only Overview

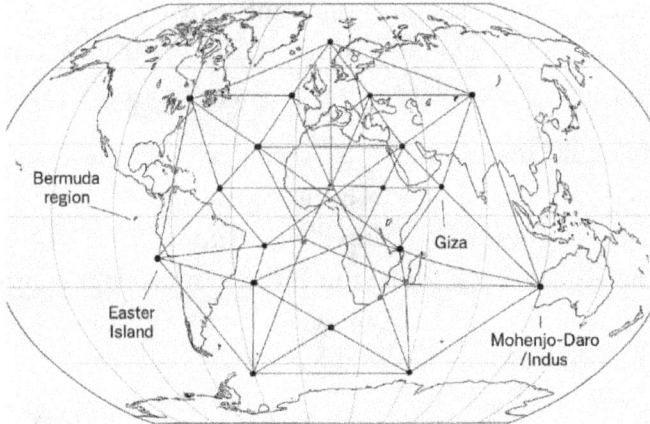

Bermuda region

Giza

Easter Island

Mohenjo-Daro /Indus

Geometry only; not a physical field map

Geometry overlay (UVG-120)

What advocates point to

1. **Node coincidence claims:** Certain nodes align with famous "anomalous" regions (e.g., Bermuda Triangle lore) or island megaliths.

2. **Geometry as a universal:** Platonic solids appear across cultures; a global polyhedral net is aesthetically and symbolically compelling, and geodesic thinking (Fuller) is proven engineering.

3. **Map echoes:** The authors argue that some medieval charts can be "read" within their triangle framework.

What critics—and cautious analysts—note

- **Projection dependence:** Where lines land depends on which world projection you use when you eyeball "matches." Replot on a globe or in great-circle math; some fits evaporate. (Even sympathetic grid enthusiasts warn about projection artifacts.)

- **Selection effects:** With thousands of significant places on Earth, some will lie near any chosen lattice. Without pre-registered criteria and proper null models, "hits" don't carry weight. Archaeologists and statisticians have made exactly this point for straight-line alignments (leys) on regional maps.

- **Evidence standards:** Much writing on grids lives outside peer-reviewed geophysics. Even open-minded archaeologists separate measurable site physics from claims of a global energetic web. (The **Dragon Project** ran radiation, ultrasound, and magnetic checks at stone circles; results showed some local anomalies but no universal pattern.)

Do Monuments "Snap" to Grids? A Practical, Testable Workflow

Let's say you want to know whether a region's megaliths relate to a large-scale net—or whether their pattern is equally explainable by visibility, terrain, water sources, astronomy, and cultural roads. Here is a sober protocol you can use (and that readers can replicate):

1. **Assemble clean datasets:**

How to evaluate a grid claim: (1) Use a globe or great-circle math, not a flat projection; (2) pre-register which sites 'count' before plotting; (3) run Monte Carlo simulations to assess chance; (4) publish coordinates, code, and thresholds.

- o Site centroids (lat/long) with date ranges and feature types (circle, cairn, temple).

- o Environmental layers: elevation, slope, bedrock (quartz content if available), faults, streams, soil conductivity.

- o Cultural layers: known procession paths, intervisibility lines, archaeoastronomy bearings.

2. **Define the grid a priori:**

   - o If testing **Becker–Hagens**, compute node and edge coordinates on a sphere (not on a 2D projection).

   - o Alternatively, define **great circles** by astronomical azimuths or constant bearings fixed *before* you look at the data.

3. **Statistical tests:**

   - o **Nearest-node analysis:** Compare observed site-to-node distances against thousands of randomized site distributions constrained to the same region and environmental mask.

   - o **Edge alignment:** For each site axis (avenue, alignment, sightline), test angular deviation from the nearest grid edge versus randomized orientations.

   - o **Environmental controls:** Run multivariate models to see whether geology/topography explains siting better than grid proximity.

4. **Instrumental checks on a subset:**

   - o Magnetometer, EM conductivity, passive VLF, and ionizing radiation logs through 24-hour cycles to catch diurnal telluric variations.

o   If quartzic bedrock is present, repeat during periods of microseismic activity to probe piezoelectric spiking.

This approach neither assumes a grid nor dismisses it. It lets the data talk.

SPATIAL GRID VALIDATION
— 6-STEP FLOW
HOW TO TEST A GRID

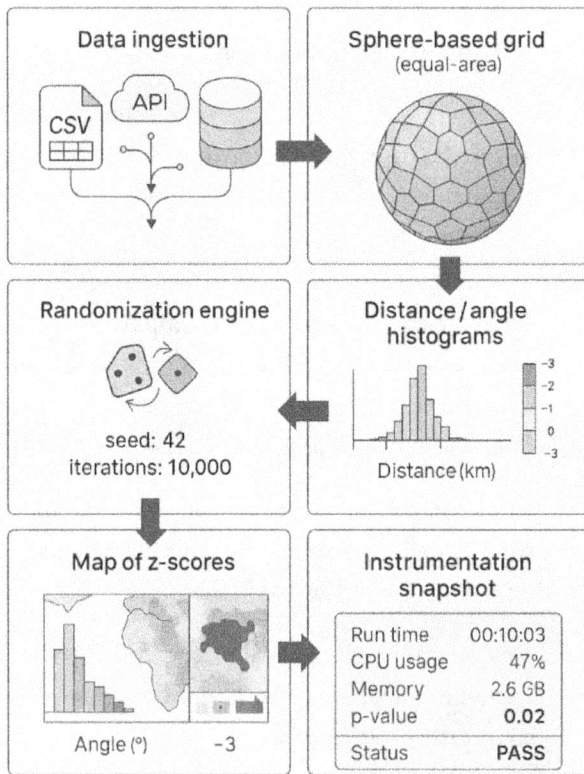

Style benchmarked to Nature/PNAS figure: clean, white, sans-sensif jeew

Where the Megaliths Themselves Nudge the Story

A few touchpoints help frame the global-grid conversation without overreaching:

- **Nazca's lines** are truly massive, precise ground graphics whose purpose remains debated—astronomical mapping, processional routes, ritualized sky-mirroring—but not literal "runways." Some studies proposed stellar alignments (e.g., Orion); others emphasize ritual pathways. Whatever the function, Nazca is hard evidence of large-scale linear planning and technical execution.

- **"Ancient memory" in mapping** resurfaces in multiple sources, arguing that some charts preserve deep-time coastlines and accurate longitudes, implying older survey and math traditions—ideas that, if true even in part, would support the *possibility* that large-scale geodesy existed earlier than we think.

- **European megalithic systems** have been argued to show continental-scale planning and linked ovals/roads (Avebury–Stonehenge corridors and beyond), reinforcing the concept of "macro-design." While controversial, this line of inquiry keeps the door open to civilizational-level projects in deep

A Balanced Reading of the Evidence

- **What's solid:**
  - Real geophysics (tellurics, piezoelectric effects) *can* produce local anomalies.

- Some monuments demonstrably encode geometry and astronomy. (Debate focuses on scope, not existence.)

- Cartographic anomalies exist on certain historical maps; whether they reflect lost sources remains debated, but the geometry on the parchment is measurable.

- **What's plausible but unproven:**

  - That ancient planners intentionally targeted electromagnetic "hot spots" because they perceived physiological or ritual effects. Small-scale studies show mixed results; bigger controlled campaigns are rare.

  - That a **single** global grid governs sacred site placement worldwide. To test this, we need preregistered methods and public data.

- **Where skepticism is healthy:**

  - Projection-based map "hits," post-hoc site selection, and numerology-driven fits. The right response isn't dismissal; it's better statistics.

How to Read the Landscape Like a Grid-Wise Researcher

When you travel or research for *Forbidden Maps*, treat every "charged" place as a lab:

Nazca is not proof of a global grid; it is proof that ancient engineers executed mile-scale straight forms with intention. That's the mindset you need for any planetary framework.

1. **Look down and in:** bedrock type, quartz content, fault proximity, groundwater.

2. **Look up and out:** sightlines, skyline notches, solar/lunar risings, far-hill intervisibility.

3. **Log the air:** simple magnetometer and Geiger logs over 24 hours, plus VLF radio noise; note weather and solar activity.

4. **Map with discipline:** if you're testing an alignment, pre-declare your points and allowable tolerance; if you're testing a grid, publish the code that generated it.

This approach honors the mystery *and* the method.

Becker–Hagens Grid: A Clear, Critical "How-To" Mini-Study

If you want to include a rigorous sidebar in your chapter, here's a compact protocol you can actually run:

- **Construct** the UVG-120 on a sphere (hexakis icosahedron). Document the exact transform and node list.

- **Select** a global set of ancient "first-order" sites *a priori* (e.g., UNESCO prehistoric monuments older than 1500 BCE), excluding anything added after you see the grid.

- **Test 1 (nearest-node):** Compute distances from each site to the nearest node. Compare with 10,000 randomized site sets constrained to landmasks.

- **Test 2 (edge bearings):** For sites with documented principal axes (temple causeway, avenue), compute angular deviation to nearest grid edge; compare with randomized azimuths.

- **Report** effect sizes, p-values, and which sites drive results. Publish a negative outcome with the same enthusiasm as a positive one.

A global grid that survives robust testing would force a rewrite of our past. It would imply deep-time geodesy, long-range surveying, and perhaps an energetic sensibility that guided sacred architecture. But even if the Becker–Hagens mesh (or any mesh) ends up as an elegant metaphor rather than a measurable scaffold, the work isn't wasted. You still gain:

- A sharper method for separating **pattern from pareidolia**.

- A disciplined way to test **ancient maps' anomalies** against real coastlines and lat/long math.

- A renewed respect for how much **geometry and sky** are wired into ancient sites—something Nazca, Avebury–Stonehenge, and many others already teach us.

And you leave readers as co-discoverers, not spectators: *here's what to look for, here's how to test it, here's where we don't know yet.*

A model that survives preregistration and blind tests becomes interesting science. A model that only works after you adjust the map projection and hand-pick sites is storytelling.

Chapter 6

Sacred Geography and Hidden Knowledge

Let's start with the puzzle that refuses to go away. From Stonehenge to Giza, Chaco Canyon to Angkor, Newgrange to Teotihuacan, major temples, pyramids, and shrines are not dropped randomly on the landscape. They face precise directions, frame seasonal sunrises and sunsets, echo star paths, and often interlock with surrounding hills, rivers, and horizon notches like gears in a clock. Across cultures, eras, and continents, builders made the sky part of the blueprint and the land a stage for celestial events. The result is a living calendar carved in stone.

Why? Because sacred architecture wasn't just about walls and roofs. It was about time, power, memory, and navigation—cosmic alignment as social order. When you grasp that, "sacred geography" stops being a romantic phrase and becomes a working hypothesis: people anchored meaning to place using astronomy, and they encoded that knowledge in alignments, distances, and sightlines that still function if you know how to look.

Sacred sites often double as instruments—architecture that measures time (solstices, equinoxes, lunar standstills) and space (cardinal directions, local horizon targets). The building is the device.

How civilizations were taught with sunlight

One very practical reason to align with stars and solstices is calendar control. Farmers, priests, and rulers needed shared, predictable markers to coordinate planting, taxes, migration, feasts, and law. The sun solves that, but only if you make it visible at the right moment.

A winter-solstice shaft of light striking the back wall of a chamber is not decoration—it's a "Now" button. It tells everyone, from the elite to the last person at the crowd's edge, that the year has turned. Alignments unify a population around a schedule without literacy or synchronized clocks. The spectacle delivers authority and certainty at once.

Solar alignments also help regulate religious drama. The narrow beam that hits an altar for ten minutes a year isn't just optics; it's choreography set by the cosmos. That choreography legitimizes priestly roles ("we make the sun enter the temple") and royal claims ("the heavens obey our order"), and it sets expectations for pilgrims ("come when the light appears").

Solstices are the most reliable "anchors" in the year. Their sunrise/sunset azimuths change slowly from day to day, creating a broad "sweet spot" for architecture to catch them. This is why solstice alignments are common and durable across centuries.

Why stars got a vote in the floor plan

Stars deliver a different kind of value: identity and ancestry. A constellation rising in a temple's line of sight ties the community to myth—sky heroes, founders, ancestral animals. The stellar calendar also marks longer cycles. The nightly shift in a star's rising time across seasons is a teaching aid; the slow drift of star positions over centuries (precession) is a cultural time capsule. When builders fix architecture to a stellar event, they embed a date range. Later generations can "read" that choice, consciously or not, as origin, legitimacy, or loss.

Stellar targeting also supports navigation. If your inland capital "speaks" the same star language as coastal wayfinding, your empire shares a compass. Sacred buildings become training hubs for astronomer-priests who double as surveyors and navigators. The star school is literally in the temple.

The land itself is scripture.

Sacred geography isn't only about angles; it's about placement. Mountain spires, volcano cones, river bends, island silhouettes, and horizon saddles become markers for the sky clock. If a summer solstice sunset drops into a V-shaped notch between two hills exactly as seen from a temple, the "instrument" includes the hills. You can't move the temple without losing the reading. That dependence is the point: it

Stars vs. sun—what's harder

The sun's calendar is easy to "catch" with architecture; star alignments demand more precision. You must know the local horizon, latitude, and which star or asterism you're targeting. Precession moves star positions ~1° every 72 years, so old stellar alignments age out unless you track that drift.

binds human design to the specific terrain; the land becomes part of the liturgy.

Cultures labeled this bond as an axis mundi—world axis—through trees, pillars, obelisks, and pyramids. The vertical form negotiates between earth and sky, while the ground plan ties the layout to cardinal directions, rivers, and sacred roads. As people walk the processional routes, they are literally tracing a cosmogram.

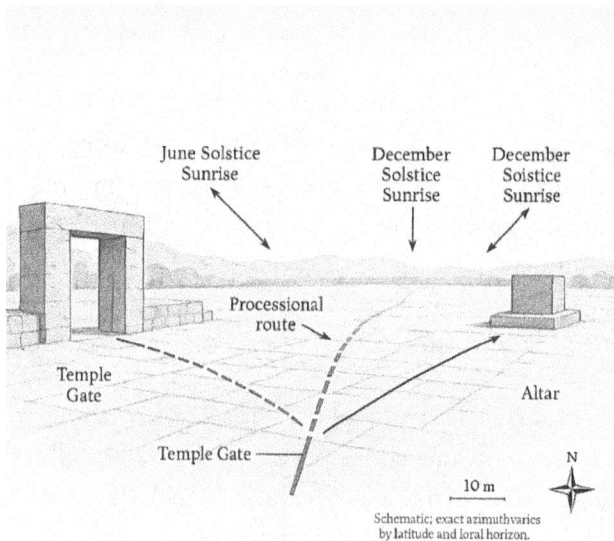

Temple Courtyard–Solstice Alignments & Processl Route
(Schematic panoram: Schematic Panorama)

Why temples, pyramids, and shrines align with stars and solstices

Let's go deeper than "because it looks cool" or "ancient people liked the sun."

1) Calendrical governance

A temple that hits the solstice sends a time signal across society. It synchronizes tax collection, agricultural labor, pilgrimage dates, and the legal year. Solar-aligned architecture is civil infrastructure in stone.

2) Ritual optics and social cohesion

A rare light effect—beam, dagger, halo—creates shared awe. This is ritual technology: predictable enough to stage, rare enough to feel miraculous. It pulls a population into a single emotional frame, reinforcing cooperation.

3) Legitimacy and cosmic charter

Alignments encode origin stories. If your founding myth claims descent from a star ancestor or a solar hero, the building "proves" the myth yearly. Power draws authority from the sky: "We rule under this sign, and the sign appears on cue."

4) Education and knowledge management

Temples function as observatories. Apprentices learn to watch first and explain second. The resulting knowledge—azimuths, declinations, and horizon altitudes—feeds surveying, road-building, plotting fields, and navigation.

5) Engineering and craft precision

Chasing alignments forces mastery over straight lines, level planes, right angles, and long baselines. If your culture can hold an azimuth across hundreds of meters, you can lay out cities, aqueducts, and causeways with confidence.

6) Pilgrimage economics

Sites that "perform" at solstices draw pilgrims—goods, offerings, labor. Alignments become an economy of attention long before tourism had a name.

7) Defense of memory

Alignments survive illiteracy and regime change. Even if texts burn, the building still teaches. Light, shadow, and sightlines carry the syllabus forward.

The connection between geography, astronomy, and myth

Myth is not a loose story thrown over random ground. It is a data carrier. Geography supplies the canvas, astronomy provides the clock, and myth ties them into a narrative people will remember. Here's how the layers stack.

How to spot a cosmogram

The six variables you must get right

Latitude; true north (not magnetic); local horizon altitude; refraction near the horizon; epoch (for stellar targets— precession); and the building's line-of-sight free of later obstructions. Miss one and your "alignment" may be a coincidence.

Cardinal order into social order.

Four directions become four quarters of a city, four gates, four clans, four colors. The ruler sits at the center—the axis mundi. The palace or main pyramid claims the crossing point of heaven's dome and earth's grid. In a single gesture, astronomy becomes political geometry.

Sky animals and earthly terrain.

If a culture traces a jaguar or an eagle in the stars, that animal is going to show up in art, names, and place-choice—ravines shaped like claws, peaks nicknamed with beaks, roads that follow "spines." The sky map informs toponyms and the paths people prefer.

Cycling gods, cycling seasons.

Sun gods age, die, and return. Lunar goddesses wax and wane. A solstice light entering a sacred chamber is literally a rebirth scene; equinoxes frame balance. Festivals sit on those moments to cement cosmology to the calendar.

Floods, fires, and resets.

Many traditions remember destructive cycles. Whether rooted in real regional disasters or folded into wider sky narratives, those memories are mapped onto specific mountains (shelter peaks), caves (rebirth), and islands (remnants). Pilgrimage routes often "re-enact" the survival journey—up a stair, through a gate, into a chamber, out to light.

Hero paths as survey lines.

A founder's legendary march from one sacred hill to another often follows a straight line on a map—sometimes suspiciously straight. Those "paths" can hide survey baselines, processional roads, or the skeleton of an urban grid.

MYTH MAP:
SUN-HERO'S YEARLY CIRCUIT

Solstice & Equinox Stops overlayed on temples, hills, and river crossings

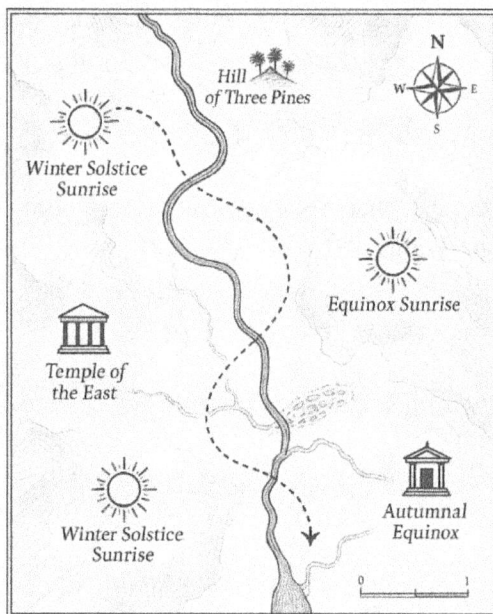

A field guide to reading sacred alignments (without fooling yourself)

You can do this responsibly with a compass, a clinometer app, a decent topographic map, and care.

Step 1: Establish true north.

Don't trust magnetic north. Correct for magnetic declination or use solar methods (e.g., shadow stick at local noon). Mark the site's principal axis in degrees from true north (azimuth).

Step 2: Record the horizon.

Measure the altitude in degrees of the visible horizon at the azimuths of interest. A hill that's 2° high will shift your target by about two days near solstices; near the equator, more. Always include the horizon in your notes.

Step 3: Identify candidate events.

Check whether the azimuths correspond to solstices, equinoxes, cross-quarter days, lunar standstills, or significant star risings/settings for the site's latitude. For stars, account for precession to the era you think is relevant.

Step 4: Control for chance.

Rectangular buildings produce lots of "alignments" by accident. Run a simple test: if you rotate the plan 5° in a sketch and the "alignment" disappears while the practical orientation (to wind, slope, or street) still makes sense, caution is warranted.

Step 5: Seek the performance.

Ask if there is a plausible light/shadow effect that would be visible to a gathered audience on a predictable date. A ten-second pinhole effect in a back room with no access? Probably not the intention. A twenty-minute beam that climbs an altar on the solstice? That's a contender.

Step 6: Find the landscape partners.

Scan for notches, peaks, sea gaps, and river bends that sit under those azimuths. If the architecture "catches" an event on a distinctive landscape feature, the case strengthens.

Ley lines: straight lines, straight talk

"Ley lines" entered popular vocabulary as mysterious straight alignments linking old churches, mounds, and megaliths across the landscape. The modern term is new; the underlying impulse is not. People have long drawn meaning from straight routes—pilgrim roads, processional ways, Roman centuriation grids, royal avenues. The question is whether there's anything more than human preference and practical surveying.

Here's the balanced view.

The skeptical case says that if you scatter enough points on a map and start drawing lines, you'll find impressive alignments just by chance. Human brains love patterns, especially straight ones. Without pre-stated rules, "proof" lines can cherry-pick points and ignore near misses. Also, old sites tend to cluster along ridge lines, river corridors, and fertile boundaries—natural straight-ish flows people have used for millennia.

The alternative case argues that some alignments are too consistent and too loaded with high-status sites to be random. Proponents extend the idea to "earth energies"—subtle fields sometimes linked to groundwater flow or geological faults—claiming sensitives can detect them and that animals and plants respond.

A sober middle recognizes that ancient surveying excellence is real. Long, ruler-straight alignments are humanly achievable with simple tools (plumb bobs, sighting poles, cords) and astronomical references. Processional causeways and pilgrimage roads intentionally connect shrines. City grids extend cardinal lines for ideological reasons. In some regions, groundwater and fault lines do correlate with settlement density (water is life, and fractures store water), which can incidentally create map patterns that look "energetic."

Case Study: The Nazca Lines and celestial mapping

Desert, hardpan soil, thousands of lines, trapezoids, spirals, and animal figures etched by removing dark stones to reveal lighter ground. The

Nazca Pampa is both overwhelming and deceptively simple. The raw recipe—draw a line and clear the surface—allows fast, large-scale execution. The tough part is interpretation. Are these a star chart on the ground, a ritual stage, a water cult, or all of the above?

Let's structure the analysis.

The dataset

Nazca geoglyphs include straight lines kilometers long, broad trapezoids with flared ends, zigzags, spirals, and biomorphs (monkey, hummingbird, spider, whales, plants). The figures avoid overlapping each other's core spaces but often cross ordinary lines. Many lines aim at horizon points; some appear to converge on vantage hills. The culture flourished roughly 200 BCE–700 CE.

Mainstream interpretations

Ritual pathways and processional grounds.

The long lines and trapezoids are excellent for walking—rituals in motion. Groups could process along them during festivals, sometimes reversing course to "charge" a space by movement. The biomorphs function as emblems or mythic beings—destinations, not highways.

> **Constraints that matter at Nazca**
>
> *Hyper-arid climate locks the lines in place for centuries; flat plains allow long sightlines; nearby low hills provide vantage points; the local horizon is clean—excellent for astronomical bearings.*

Vantage hills serve as viewing stations for orchestrating and supervising ceremonies.

Water and fertility cults.

In a fragile desert economy, water is survival. Nazca engineered underground aqueducts (puquios) with spiral access holes. Spirals on the pampa echo this technology and may invoke or honor the same principle—coaxing water from the earth. Ritual walking, offerings, and geoglyphs could be directed to "wake" or thank the water.

Astronomical cues as scheduling aids.

Even within ritual and water frameworks, celestial markers help time events—"walk when the sun rises there," "offer when the moon sets there." The lines can be multipurpose: navigational during processions, calendrical when extended to the horizon.

Celestial mapping claims

Some researchers have proposed that specific lines and trapezoids target solstice and equinox sunrise/sunset points; others extend this to stellar risings. The astronomy at Nazca is complicated by the quantity of lines: with enough angles, many will hit interesting azimuths by chance. That doesn't invalidate the idea; it raises the bar for proof.

Stronger indicators include clusters of parallel lines that bracket a solstice range, sightlines that work from a known platform toward a distinctive horizon feature, and figure orientations that match animal-star myths. For example, if a trapezoid's long axis frames a winter solstice sunset within a degree, and nearby there's a standing stone or platform aligned to the same azimuth, and people could gather there, that's substantive.

Weaker indicators are isolated single lines claimed to hit a bright star without attention to local horizon altitude, epoch shifts due to precession, or refraction. If you need to jump centuries to make a star fit, caution.

A hybrid model that fits the evidence

The most sensible reading treats the pampa as a ritual instrument with calendrical settings. The lines and trapezoids provide routes and framed views; the animal figures brand sacred precincts and encode myth. Water technology and iconography thread through it all. Astronomical targeting appears where it is handy—solstice sunset down a trapezoid, equinox sunrise along a straight line from a hilltop—but the site is not a full-sky "map" in any modern sense.

This hybrid model matches what we know from other cultures: when

How to test a Nazca alignment

Pick one trapezoid. Measure both axes' azimuths; record horizon altitude in those directions. Compare to solstice/equinox azimuths for the site's latitude; adjust ±0.5–1.0° for horizon elevation. If it hits, ask: is there a nearby platform oriented the same way? Is the view uncluttered? Is there a plausible festival date tied to that event?

people invest huge effort into landscape-scale lines, they do it to move bodies, signal times, and invoke powers tied to both the sky and the ground. You don't need only one motive when three reinforce each other.

Hidden knowledge: what was guarded, and how

The phrase "hidden knowledge" often gets inflated into secret science or visitors from elsewhere. Let's ground it. What kinds of knowledge were actually hard to get and worth guarding?

Long cycles and their leverage.

Tracking a lunar standstill (~18.6-year cycle), predicting eclipses, or accounting for precession across generations demands more than casual observation. Cultures that did this had structured observation programs and teaching pipelines. They guarded the methods because calendar mastery equals political power.

Surveying that scales.

Holding a straight baseline across valleys, transferring cardinal directions accurately to distant sites, and reconciling local terrain with celestial targets takes craft. Surveyors are valuable; they train within institutions (temples, royal schools). The "secret" isn't mysterious energy—it's accurate fieldwork and good instruments.

Time-linked ritual technology.

If your main shrine "activates" at a moment no one else can anticipate, you own the public clock. The schematics of shafts, apertures, and chambers that produce those effects are guarded like state assets.

Encoded maps.

Some sacred landscapes embed distances, ratios, or bearings that double as memory aids for travel or boundaries. A pilgrim route might also be the fiscal limit of a capital's reach. Insiders know both readings.

Taboos as encryption.

Not all secrecy is technical. Limiting who can climb a temple stair or read a floor plan keeps the performance reliable and the hierarchy clear. The fewer hands in the mechanism, the less chance someone breaks the clock.

Worked examples across cultures

Stonehenge (England).

The monument's axis points to the summer solstice sunrise and winter solstice sunset. Massive efforts to erect and refine the arrangement anchor a simple but central message: "We know the year." The avenue linking Stonehenge to the River Avon aligns with solstitial directions, connecting sky, stone, and water.

Newgrange (Ireland).

A narrow roofbox admits winter solstice sunrise for a few minutes, illuminating the deep chamber. It's precision within a forgiving target (solstices move little day-to-day). The performance is so good that even today crowds gather—and that's the point: it always was an event for crowds.

Teotihuacan (Mexico).

The urban grid is rotated ~15.5° off true north. Debates continue over the "why," but several horizon and star ideas compete: sunsets tied to agricultural dates, or a city plan keyed to a specific calendrical count. The key is the deliberate choice: cardinal directions were available; a different orientation was preferred for layered reasons.

Machu Picchu (Peru).

Intihuatana ("hitching post of the sun") and careful window placements frame solstitial sun paths and bright stars peeking over peaks. The surrounding horizon—a theater of distinctive mountains—is part of the instrument.

Angkor Wat (Cambodia).

The complex faces west; the moat and towers produce equinox/solstice sunrise/sunset compositions from key vantage lines. Beyond day-to-day sun, the bas-relief program encodes cosmology (churning of the ocean of milk) that ties celestial order to royal mandate.

Chaco Canyon (USA).

The Sun Dagger at Fajada Butte—a narrow light crossing a spiral at solstices and equinoxes—suggests skilled use of light and shadow on rock art. Major Great Houses align with cardinal directions and with each other across the canyon with long roads, implying survey ability and social coordination.

Egyptian temple chains (Nile Valley).

Processional alignments tie temples along the river; pylons frame horizon events; obelisks act as sun dials writ large. The river provides a north-south baseline; the sky tunes the detail.

Methods: how ancient builders nailed the angles

You don't need advanced technology to align a building precisely. You need repeatable procedures, patience, and tools that any mason could build.

Finding cardinal directions. Stick-and-shadow at local noon gives north-south; equal-altitude marks morning and afternoon azimuths to bisect east-west; star pairs that straddle the meridian help at night. Water bowls and mirrors can refine observations.

Measuring azimuths and horizons.

A simple sighting frame with a plumb bob sets a clean line. For horizon altitude, a stick with a marked angle (a basic clinometer) does the job. Repeating measurements across days improves accuracy.

Transferring lines to distance.

Sight poles on a clear day, cords to keep straightness, and intermediate markers in valleys let you hold a baseline over kilometers. A controlled error budget—small, known deviations—keeps the whole grid honest.

Correcting for local realities.

Horizon notches change apparent solar azimuths; builders either selected sites with clean views or embraced the notch as the feature to frame. Refraction near the horizon can lift the sun's apparent position; repeat observations averaged across years help settle on a robust target.

Risks and red flags in alignment hunting

Error is human; robustness is divine

Good sacred alignments build in wiggle room. Solstices give you a wide azimuth plateau; large doorways and broad corridors provide a "catcher's mitt" for the light; then a narrow niche refines the moment.

Data dredging. With enough lines, some will match something. Pre-register your test: "I will look for solstice/equinox, not every bright star over 2,000 years." Then stick to it.

Magnetic vs. true north mistakes. Compass readings without correction mislead. Always convert to true north.

Forcing precession to fit a favorite epoch. If your star match only works at 800 BCE but the site dates to 1200 CE with no rebuild evidence, you're bending reality.

Ignoring the horizon, Maps don't show the local horizon; you must stand there or use a horizon profile. A 2° hill changes everything.

Private effects Alignments that only a lone caretaker in a sealed room can see are suspect. Sacred architecture favored shared drama.

Building your own sacred-site dossier

If you're documenting a site, produce a dossier that a skeptic can read and still learn from.

1. **Site plan** with measured axes and true-north reference.

2. **Horizon profile**—sketch or photo panorama with degree marks.

3. **Observation log** for sunrise/sunset dates across seasons.

The "three-leg" test

A credible alignment stands on three legs: orientation, horizon, and context (ritual/art/route). Kick any leg away and the case falls.

4. **Cultural context notes** (myth, festival calendar, art program).

5. **Photographic evidence** of any light/shadow events, with time stamps.

6. **Error discussion**—your margins, uncertainties, and alternate readings.

If you came for "ley lines," here's the honest elevator pitch after everything we've covered. There are real power lines—just not in the fantasy sense. They are social and cognitive lines:

- **Lines of sight** that bind hills to temples to sunrise points.

- **Lines of procession** that tie neighborhoods to a civic calendar.

- **Lines of narrative** that map myth onto geography.

- **Lines of measurement** that seed engineering competence.

- **Lines of authority** that run from sky events to political legitimacy.

These are the lines empires are built on. They are strong because they enlist the sun, the stars, and the land itself as co-authors.

Pick one site near you. Measure it. Watch it through a year. Learn its horizon. You'll join a chain of observers older than history, doing the simplest and most profound thing humans do: aligning ourselves to the world we live in.

Strip away the mystique, and what remains is impressive enough. Sacred geography trains people to pay attention—to the first gleam of dawn in a notch, to the way a corridor collects light, to the feeling of a crowd turning toward the same horizon. It blends practical needs with meaning. It is conservative without being dull, innovative without being chaotic, and precise without being fragile.

The idea isn't to pretend every old stone points to Sirius, or that straight lines on a map prove a global master plan. The idea is to give due credit: many builders knew exactly what they were doing with the sky, the land, and the human mind. They left instruments behind— big ones. Some still work. Go stand in the right place at the right time, and you'll see the calendar turn the way they intended.

The take-home

The Earth doesn't need invisible energy to feel alive. Once people braid sky, land, and story, the landscape becomes a circuit anyway—powered by human coordination and memory.

Part III: Suppressed Cartography and Forbidden Knowledge
Chapter 7: Maps the Authorities Ignored

You'd think maps would be the last place to hide a controversy. They're supposed to be neutral, right? Coastlines, bearings, ports, the practical stuff of getting home alive. Yet tucked into archives, stitched onto vellum, and scrawled in cramped marginalia are maps that don't behave. They suggest coasts where there "shouldn't" be coasts, islands that winked in and out of existence, and geometric systems it supposedly took us centuries to rediscover. This chapter is about those maps—how they were found, why they were sidelined, and what happened when military and intelligence professionals quietly took a look for themselves.

A quick orientation

Three strands twist through this chapter:

1. Why controversial maps have been left out of mainstream history narratives.

The Promise and Problem of Old Maps

- *Old charts often compile even older sources. That makes them powerful—and error-prone.*
- *Portolan designs can mimic precision without modern latitude/longitude grids.*
- *To read them responsibly, you must separate projection artifacts from genuine geography.*

---

2. What declassified military and intelligence material actually says about old charts.

3. A case study in "phantom islands"—landforms that haunted official charts for centuries and then vanished.

We'll keep the tone grounded. Where claims go too far, you'll see the conventional counter-arguments. Where the mainstream story is thin, you'll see the anomalies that won't quit.

Why controversial maps are excluded from history books

Let's start with the human part. University historians, museum curators, school textbook panels—none of these groups are set up to reward "could be." They're tasked with teaching what stands up under replication and hard evidence. That's not censorship; it's quality control. But that filter can also sideline awkward data that's hard to classify: a fragmentary chart, copied from older charts, drawn on an unknown projection, with annotations in a script that's only partially understood.

Take an Ottoman world chart compiled in 1513 and rediscovered centuries later. It's a composite—its own drafter says so—built from "about twenty" old charts and mappae mundi, plus newer Portuguese and Spanish sources. That composite nature is precisely why it fascinates researchers…and why it makes traditional historians wary. A composite can store genuinely ancient data—but it can also propagate ancient mistakes.

When researchers in the twentieth century tried to reverse-engineer the projection and grid behind that 1513 fragment, they found an underlying geometry consistent with a sophisticated portolan framework—one that could, with care, be translated into latitude-longitude terms. That mathematical reverse-engineering lit a fuse: if the grid was real and consistent, maybe some of the coastal depictions were, too.

Then the fight started. On one side: "These southern landmasses match modern polar geography with uncanny fidelity—how could sixteenth-century compilers do that unless they had older, better sources?" On the other side: "What you're calling Antarctica is just a stylized Terra Australis, or a misaligned extension of South America— or simply the cartographer squeezing land to fit parchment." Skeptical historians have also stressed that later surveys overturned some of the mid-century seismic interpretations originally used to support an ice-free coastline claim; in their view, the resemblance is tenuous and the 'Antarctica' identification fails basic checks on South America's missing miles.

So why do standard histories skip these debates or relegate them to endnotes?

Five practical reasons:

- **Chain-of-custody headaches.** Many old charts are copies of copies, with missing halves and unknown maprooms in their pedigree. Without a clean provenance, bold claims stall.

- **Projection illusions.** If you misread a portolan or conflate wind-rose rhumb networks with lat-long grids, you can "discover" accuracy that's really geometry. Historians prefer errors they understand over precision they can't audit.

- **Curricular inertia.** High school and college surveys chase consensus; they're already overfull. Edge cases get trimmed.

- **Risk of sensationalism.** One sensational chapter can taint a whole textbook. Committees avoid it.

- **Moving scientific baselines.** Interpretations built on early seismic maps or preliminary polar data can age badly, and textbook writers don't want a boomerang.

Declassified military & intelligence interest in ancient cartography

It isn't just enthusiasts who've weighed these charts. During the Cold War—when cartographic accuracy was a national security asset—professional analysts quietly reviewed some of the best-known "anomalous" maps.

What the U.S. Air Force said (and didn't say)

A cartographic unit within a Strategic Air Command reconnaissance squadron examined the 1513 chart and a sixteenth-century world map noted for its elaborate southern landmass. In formal correspondence, they assessed specific southern coastal features and, comparing them with then-recent seismic profiles, wrote that the interpretation aligning portions of the chart's southern coast with coastal Antarctica seemed "reasonable," even "the most logical." They underlined how astonishing it would be if that coastal detail pre-dated the modern ice

How to Vet a Controversial Map

- *Identify its projection before you judge its coastline.*
- *Reconstruct its compilation method (single survey vs. multi-source collage).*
- *Audit the error budget: where are errors clustered, and why?*
- *Compare against independent datasets (bathymetry, sub-glacial topography, and later hydrography).*
- *Track provenance and marginal notes—they often reveal sources and motives.*

cap—and, crucially, they admitted they had no idea how to reconcile such data with sixteenth-century geography. That cautious, technical language matters: it isn't hyperbole; it's an analyst's way of saying, "We can't explain this with what we know."

A subsequent, more expansive summary from the same Air Force cartographic office praised the projection reconstruction, suggested that both maps drew on accurate source materials, and proposed that the southern landmass might reflect a time when major waterways and coasts were relatively free of ice. The memo even speculated about advanced projections in the source material. Again—the tone is analytic, not evangelical. They're evaluating geometry and fit, not arguing a sweeping historical thesis.

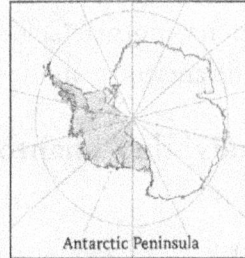

DECLASSIFIED
DEPARTMENT OF ▮▮▮▮
MEMORANDUM
TO ▮▮▮▮
FROM: ▮▮▮▮
SUBJECT: COASTAL RECONNAISSANCE SUMMARY

May 14, 1971

DECLASSIFIED
FOIA ▮▮ HO.
RELEASED IN PR.

Joint coastal reconnaissance activities with ▮▮▮▮ forces ▮▮▮▮, investigations of inlets, bays, and channels in the coastal segment referenced in attached diagram. Key·antiteal.

Aerial photography and ghipb'aboard observations survey purposes identified hey specific ice conditions, navigability, and potential anchorage sites were identifd.

I recommend this this findings be valuable for... Suggest further missionton next year to survey sectors of the constiine pending available resources.

Antarctic Peninsula

Suggested title:
Declassified Memo & Polar
Inset -Antarctic Peninsula
Segment

What the State Department files show

Decades before those Air Force memos, diplomatic correspondence documented the rediscovery and early translations of the 1513 chart, including Ottoman-Turkish marginal notes about sources (older charts, mappae mundi, and Iberian maps). Those letters are dry, administrative—even better as historical anchors. They verify that officials took the map seriously enough to request analyses, photographs, and translations through embassies and ministries.

What the CIA files tell us (and don't)

The Agency's cartography center produced thousands of maps over the decades, many now declassified. That release shows institutional respect for cartographic craft at the heart of intelligence work—less about ancient mysteries, more about accurate modern mapping.

As for "ancient cartography," the CIA's FOIA reading room includes abstracts monitoring foreign scientific articles that debated these very charts—evidence that intelligence analysts tracked the public conversation (a normal Cold War practice), not that the CIA endorsed any specific ancient-map hypothesis.

Case study: the "phantom islands" that disappeared from official charts

What the Air Force Letters Actually Do

- *Confirm: the analysts found the projection work competent and some southern coastal matches persuasive.*
- *Acknowledge: the implications don't square with standard timelines.*
- *Do not claim: a lost civilization, a conspiracy, or definitive proof of ice-free Antarctica on a Renaissance map.*

Phantom islands are the perfect stress test for cartographic humility. They're landmasses that appeared on credible charts for years— sometimes centuries—and then evaporated under the pressure of better measurements. They remind us that maps are arguments as much as they are pictures.

Hy-Brasil, Antilia, and the ghosts of older shorelines

Medieval and Renaissance portolan charts are thick with islands that shouldn't exist—until you realize some may be distorted echoes of real land, mapped in a different time or from far-traveled sources, and then misplaced. Hy-Brasil, the west-of-Ireland apparition, persisted on charts long after ships failed to find it, and at least one chart kept it into the nineteenth century. Another "island," Antilia, appears prominently on a 1424 Venetian chart and stays in cartography for centuries, its shape and companions shifting names and locations as mapmakers wrestled with better data.

One modern hypothesis argues that some "mythical" Atlantic islands on early charts could be mislocated renderings of East Asian islands copied from Chinese sources that reached European workshops, then were dropped into the wrong ocean by compilers who had no Americas to anchor them. In that reading, Antilia's outline and river mouths line up intriguingly with Taiwan; a nearby island could be the Pescadores; and the cluster to the north resonates with Japan's main islands and Hokkaido. Whether you accept the full argument, it's a masterclass in how shapes migrate through copying chains—and how phantom islands can be the by-products of **real** islands, shifted by faulty frameworks.

Sandy Island: deleted in the age of satellites

In 2012, an Australian research vessel sailed through a well-charted patch of the Coral Sea, bound for a rectangular blob that appeared on global maps and even Google Earth. It found deep, empty water. The phony island—Sandy Island/Île de Sable—was quickly erased from many datasets; France had actually removed it from its hydrographic charts decades earlier. The most likely culprits were a century-old erroneous report and cascading data-inheritance across map publishers—amplified by the authority of digital maps.

Buss Island: the North Atlantic that wasn't

Reported in 1578 during a famed northern voyage, "Buss Island" haunted North Atlantic charts into the nineteenth century. As transatlantic traffic increased and soundings improved, the island shrank to a "sunken land," then vanished—likely a mix of dead-reckoning error and optical phenomena near Greenland and Baffin Island.

Phantom Islands: Three Mechanisms

- *Misplaced reality: Real Island, Wrong Ocean, or wrong coordinates after copying across projections.*
- *Optical/observational error: breakers, fog banks, pumice rafts, ice, or cloud lines misread as land.*
- *Data lag: once printed, an error "sticks" for decades until an authority removes it.*

Bermeja: the missing islet with geopolitical consequences

Bermeja, a tiny islet once shown off the Yucatán coast, quietly influenced modern maritime boundary thinking because, if real, it could have shifted exclusive economic zone claims in a hydrocarbon-rich area. Surveys in 1997 and again in 2009 found nothing at the charted site, and scholars generally treat the islet as a repeated cartographic mistake—though popular imagination still spins conspiracy theories about deliberate removal. The case is a reminder that even a speck on paper can cast a long legal shadow.

The southern puzzle: when a "terra incognita" looks too real

Return to the sixteenth-century world maps with robust southern continents. Skeptics call them fantasies—decorative Terra Australis blobs common long before Antarctica's modern discovery. Proponents argue that some editions, when you peel back projection errors and compilation seams, resolve into coastlines and river estuaries that look shockingly like what seismic surveys later revealed beneath the ice.

A famous sixteenth-century world map often cited in this debate shows mountains edging the southern coasts and river systems draining to the sea—features that one detailed comparison later set against known sub-glacial topography maps from the International Geophysical Year, noting positional resonances and plausible drainage patterns. Even friendly analysts acknowledged scaling and placement problems, and suggested the compiler had stitched together several local coastal maps with inconsistent scales and projections—remarkably accurate in places, mismatched in others. That kind of mosaic behavior is exactly what you would expect from a compilation project built from uneven, older sources.

Where mainstream skepticism bites back

A thorough modern critique points out that if you identify the southern landmass as Antarctica, the same chart appears to omit thousands of kilometers of South American coastline—an inconsistency hard to swallow. It also argues that later, better surveys of Antarctica's edge don't fully support the "ice-free" coastline claim that early proponents leaned on. In this view, the southern landmass is best explained as stylized Terra Australis or a squeezed South America, not a memory of pre-glacial mapping.

That push-and-pull is healthy. It keeps us from over-reading coincidences while forcing skeptics to explain away weirdly specific details. Both sides have homework.

Why authorities ignored these maps—and what to do differently

Reason 1: Citation traps. If a chart's most explosive claims rest on annotations ("I used 20 old maps") without those source maps surviving, academics balk. The correct response isn't dismissal; it's forensic reconstruction: projection analysis, error-field mapping, and rigorous comparison to independent datasets.

Reason 2: Collage cartography. A collage can bake in anachronisms. A fifteenth-century compiler could be copying a Roman shoreline next to a medieval port list next to a colleague's contemporary sketch. Sorting that out is hard, but not impossible if you treat each coastline segment as its own dataset.

Reason 3: Incentives. A historian who stakes a career on a controversial reinterpretation of a single map risks a lot. Institutions tend to reward cautious syntheses over high-variance ideas.

What to do differently now:

Mosaic Maps 101

- *Tell-tale sign #1: Longitude errors diverge in different sectors (eastward in one region, westward in another), implying stitched local charts.*
- *Tell-tale sign #2: Some coastal segments align cleanly only when rotated or rescaled independently.*
- *Tell-tale sign #3: Interior is vague while coastal drainage is oddly specific—consistent with mariners' priorities*

- **Open the lab door.** Publish the GIS and projection reconstructions so critics can rerun them.

- **Bake in skepticism.** Demand that "matches" survive blind tests and alternative projections.

- **Invite hostile audits.** A conclusion that survives adversarial review earns its place.

- **Teach the ambiguity.** History classes can spare two pages for how we handle anomalies; that's good epistemology for students.

A Simple Scoring Rubric for Old-Map Claims

Score 0–5 on:
(a) projection clarity, (b) source transparency, (c) reproducibility, (d) cross-dataset agreement, (e) error-field coherence. Claims that don't crack 15/25 belong in a footnote, not a headline.

DISPUTED COASTLINE EVIDENCE RUBRIC

Segment: _____

Coordinates: _____

SOURCE DETAILS

Source type	Date	Datum/Tide	Scale/Resolution	Citation
A. Hydrographic	1996	MHW	1:50 000	Commi
B. Gonteller Centrhot 2	Sentinl-2	HWL	HWL	Exa. 55

SCORING SLIDERS

Temporal Freshness	3/5 ━━━━○═══▷	5/5
Spatial Resolution	2/5 ━━━━━━○▷	4/5
Georeferencing Quality	4/5 ━━━━━○▷	3/5
Tidal/Datum Suitability	4/5 ━━━━━○▷	3/5
Method Transparency	4/5 ━━━━━○▷	4/5
Cross-source Agreement	2/5 ━━━━━○▷	4/5
Legal Baseline Relevance	5/5 ━━━○══▷	3/5
Exemple:	5/5 ━━━━━━━▷	3/3

Benchmarks common coastal mapping standards: shoreline proxy, datum, uncertainty, UNCLOS baseline.

The thrill and the caution: what these maps might still be telling us

Even if you put strict brackets around the boldest interpretations, the dossier leaves us with three sober insights:

1. **Medieval/Renaissance compilers were better data engineers than we give them credit for.** They stitched heterogeneous sources into workable navigational tools and used geometric frameworks that repay modern reverse-engineering.

2. **Government analysts didn't laugh these maps off.** They did careful projection checks, compared coastlines to then-new

geophysical data, and wrote measured notes acknowledging anomalies they couldn't explain with standard narratives.

3. **Phantom islands prove cartography evolves.** If an island can hold its ground for centuries and then evaporate under a research vessel's keel, we should be humble about both our ancestors' errors and their occasional, baffling accuracies.

Practical reading guide: how a modern researcher should approach "forbidden" maps

Start with the **object**, not the legend. What material is it on? What's missing? Are there folding creases that could distort measurements? Next, identify the **projection** and **rhumb networks**. Do not compare a portolan collage directly to a lat-long globe without reconstructing its construction geometry. Work coast by coast, segment by segment, and calculate the **direction and magnitude of errors** by region. Diverging error directions in different sectors often indicate a stitched compilation—e.g., eastward average error in one coastal arc, westward in another.

Then, read the **marginal notes** carefully. When a compiler tells you he used older charts, mappae mundi, Iberian sources, and an earlier transatlantic chart, believe the process even if the specifics are lost. That's not proof of antiquity; it's a roadmap for where to look for analogues.

Finally, **triangulate** with independent datasets: sub-glacial topography, bathymetry, and tectonic features. Where a coastline sketch happens to ride along the edge of a deep trough or across a mountain front we only later surveyed, flag it—it could still be a coincidence, but it's the kind of coincidence worth trying to falsify.

A balanced synthesis: what belongs in the history books

What should a fair history survey say today?

- **Portolan brilliance.** Medieval and Renaissance mariners left an engineering legacy in rhumbline networks and practical coastal compilations that deserve more airtime.

- **Documented Cold War curiosity.** Military cartographers evaluated several controversial maps with professional sobriety; their memos noted striking correspondences and unresolved questions. That's part of the story of modern cartography, not a footnote to it.

- **The phantom-island cautionary tale.** From Hy-Brasil and Antilia's cartographic peregrinations to Sandy Island's twenty-first-century deletion, the atlas is a living document—capable of error and correction at any time.

- **Competing interpretations.** When a southern landmass on a sixteenth-century map "looks right," mainstream critiques must be presented alongside the anomaly—missing South American coastline, projection confusions, and updated polar data that undercut earlier claims.

Map Myths & Methods:
A Four-Panel Micro–Gallery

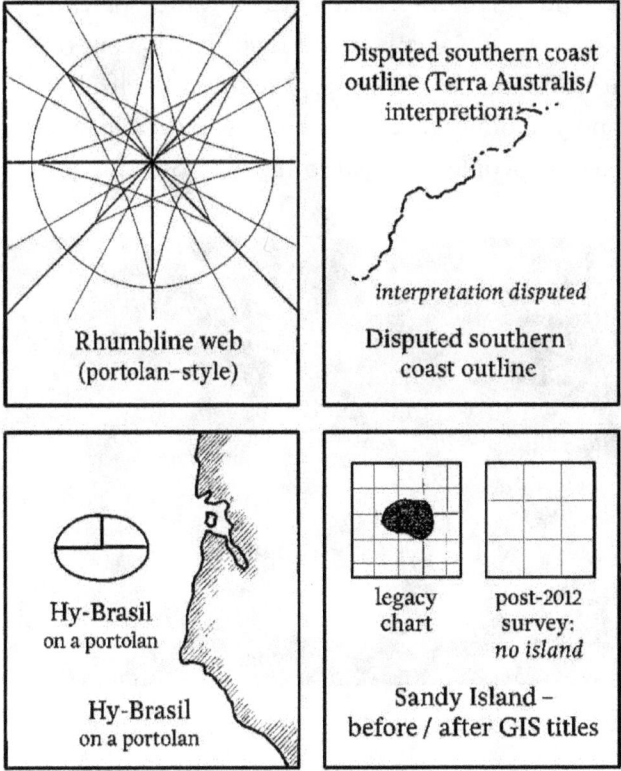

Rhumbline web (portolan–style)	Disputed southern coast outline (Terra Australis/ interpretion.) *interpretation disputed* Disputed southern coast outline
Hy-Brasil *on a portolan* Hy–Brasil *on a portolan*	legacy chart post-2012 survey: *no island* Sandy Island – before / after GIS titles

Map-Myths & Methods
A Four-Panel Micro-Gallery

Closing the distance between mystery and method

This field thrives when two instincts collaborate: curiosity and discipline. Curiosity asks: "What if this really does capture a piece of geography out of time?" Discipline replies: "Prove it against modern data and hostile tests." That's not a buzzkill; that's how remarkable insights survive.

So, were there maps the authorities ignored? Yes—because authorities aren't set up to prize ambiguity. Were there maps the authorities examined? Also, yes—and the declassified record shows they sometimes came away impressed, puzzled, and professionally non-committal. Did phantom islands really waste space on charts for hundreds of years? Absolutely—and their deletions in the satellite era prove that cartography never stops editing itself. Put all of that together, and you don't get a conspiracy; you get a living atlas, written and rewritten by people who had to steer by something.

> *Early modern world maps sometimes preserve fragments of older geographic knowledge through complex compilations on non-modern projections. Military cartographers in the mid-twentieth century acknowledged intriguing correspondences between some sixteenth-century southern coastlines and modern polar data, while mainstream historians typically attribute the same features to projection error, Terra Australis conventions, or misassembled South American coasts. Meanwhile, the phenomenon of 'phantom islands'—from Hy-Brasil to Sandy Island—demonstrates how cartographic errors can persist for centuries before correction.*

Chapter 8

The Vatican's Hidden Atlases

You're about to walk into a locked room most people only speak about in whispers: the place where sacred maps and state maps once overlapped, and where the custodians of Christendom's memory learned to keep some geographies quiet. We'll follow the paper trail—sometimes interrupted, sometimes scorched—from the age of manuscript mappaemundi to the era of missionary world charts, and we'll watch how "holy" space and "useful" space were braided together, guarded, and, at times, weaponized.

Along the way, we'll keep our footing on two paths at once. On one, the mainstream view of library practices, archival jargon, and the Jesuit cartographic machine. On the other hand, the heterodox claims that certain early-modern maps encode memories of coastlines and oceans last seen in the late Ice Age, hints of a civilization behind the curtain of history. The evidence has to carry itself; I'll show you what persuades, what doesn't, and what remains unresolved.

Promise of the Chapter

- *What "hidden atlas" means in practice (and what it does not).*
- *How Church institutions curated, tagged, and sometimes withheld geographic knowledge.*
- *A close, practical case study of Jesuit map-making and why certain world charts were circulated internally rather than publicly.*
- *How to read claims about "impossible" maps without getting lost in either credulity or dismissal.*

A door with two keys

A map can be public and private at once. Public in its design—coastlines, rivers, wind roses—yet private in its annotations: the cargoes to pursue, the saints to invoke, the anchorages to keep quiet, the political lines not to cross. In Renaissance and Baroque Europe, libraries inside and around the Vatican operated with two keys: one scholarly, one administrative. The first key opened the reading desks to humanists, friars, and visiting scholars. The second key opened cupboards of sensitive material—papal diplomacy, missionary intelligence, and geographic instruments that doubled as tools of policy.

When people say "Vatican Secret Archives," they often imagine a pit of occult manuscripts. The historical reality is more prosaic and, in a way, more formidable. "Secret" (secretum) in Latin meant "private" or "personal," as in the private archive of the papacy. A "hidden atlas," in this world, was not a magical grimoire; it was a working compilation with circulation strictly controlled: drafts, corrected proofs, marginal instructions, or compiled sheets that joined theological geography (the world as a story of salvation) with navigational geography (the world as a set of actionable routes).

Secret maps in the Biblioteca Apostolica Vaticana

What a "Hidden Atlas" Looks Like

- *Composite: manuscript sheets + printed leaves pasted into a working volume.*
- *Dual annotation layers: theology (pilgrimage, relics, shrines) and policy (ports, forts, customs).*
- *Circulation marks: ownership stamps, "for internal use" notes, delivery slits where cord ties once sealed a cover.*

The Biblioteca Apostolica Vaticana (BAV) is the visible mind of Catholic Europe—a repository for codices, printed atlases, portolan charts, and missionary reports. Not every map it held (or still holds) was intended for the reading room. Some were essentially field notebooks: loose sheets bound later when a compiler returned from Lisbon or Antwerp with items to be "quietly received," tagged, and only consulted by those with cause.

You can trace the logic. Certain geographies had theological stakes: pilgrimage roads to Rome, Compostela, Jerusalem; relic itineraries; diocesan boundaries. Other geographies had strategic stakes: which river mouths could float an armed carrack; what ports had fresh water, timber, and safe holding ground; which coastal societies were open to alliance; where the line between Portuguese and Spanish claims actually bit into the world. If you had one capital that sat atop both kinds of geographies, you would guard the overlap.

There's a second thread: the idea that Renaissance and early-modern mapmakers drew on far older source charts—some of them with a precision that shouldn't have been possible before the 18th century's long-longitude breakthroughs. The argument appears in classic discussions of early portolans and famous enigmas: a 1513 Ottoman-compiled chart said to rely on earlier sources; a mid-16th century compilation that seems to sketch a deglaciated Antarctic coastline; and other oddities that hint at a pre-classical cartographic inheritance. However one judges those claims, a plausible staging ground for preserving old prototypes between antiquity and modernity was precisely here: in Christian scholarly centers, Byzantine treasuries, and later the libraries and cabinets of Italy, where fragments could be conserved, recopied, and re-projected.

Chain of custody: from Alexandria to Rome (and why that matters)

Here's the conservative through-line not everyone knows. The people who actually copied and bound maps after late antiquity were not always explorers; they were often librarians. When the great library traditions of the ancient Mediterranean unraveled, pieces survived in monastic houses and in the Eastern Roman (Byzantine) world. After the Fourth Crusade's sack of Constantinople (1204), knowledge—books, charts, and technical compilations—moved westward through Venetian hands to Italy. From there, routes of preservation are exactly what you would expect: cathedral chapters, princely cabinets, and papal repositories. That's a sensible path for how a handful of anomalous charts (or fragments) could surface—or be quietly reworked—during the Renaissance without requiring a continuous civilian seafaring tradition from the Ice Age to Columbus. It just requires good copyists and conservative custodians.

Now layer the heterodox claim on top: some early-modern charts look as if they inherit memory from maps older than classical Greece—because they exhibit composites of coasts, long shores, and relative positions that seem to depend on advanced geodesy and surveying. Proponents point to features like surprisingly good relative longitudes on a 1513 compilation and to mid-16th-century sheets that appear to trace Antarctic coasts under less ice. Skeptics counter that creative coast-matching and projection misunderstandings can manufacture illusions of precision. Both sides agree on one thing: if truly anomalous prototypes existed, medieval and early-modern Catholic and Byzantine repositories are exactly where they could have rested between "worlds."

From Alexandria to Rome:
Transmission of Cartographic Knowledge

| Alexandria Scriptorium | Byzantine Treasury | Venetian Cartolano | Papal cabinet |

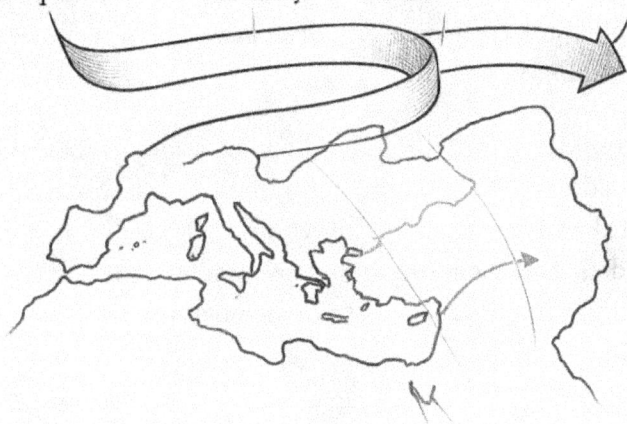

How sacred geography was controlled and guarded

A Church interested in souls was, unavoidably, interested in roads. The mechanics were simple and powerful:

Pilgrimage economies. Pilgrimage shrines—St. Peter's in Rome, St. James in Compostela, the Holy Sepulchre—sat atop networks of lodging houses, relic stations, indulgence practices, and civic privileges. Maps that emphasized or clarified those roads were good for piety, yes, but also for revenue and order. The "authorized routes"—Rome's Via Francigena, for instance—appear again and again in manuscripts and later printed guides. Keeping the official versions tidy and limiting confusion was part devotion, part logistics.

Boundary theology. Diocesan borders, metropolitan sees, and jurisdictional maps settled disputes. A bishop's authority ran along lines that had to be drawn and, if need be, redrawn. Here, "sacred geography" was literally the law.

Information triage. Missionaries sent back letters with precise geographic detail: which river passes navigable seasons, which upland valleys welcomed catechists, which chiefs carried Portuguese firearms, which plague roads to avoid. Those letters often traveled through orders' headquarters and into papal hands. Some of that content became public and edifying; some was kept in-house files—library rooms within library rooms.

Geopolitics under vows. The Catholic monarchies of Iberia guarded their state maps—the Casa da Índia in Lisbon and the Spanish Padrón Real handled master charts. Churchmen, especially members of teaching orders, had the training to copy, reconcile, and correct these charts. But their first loyalty was to their superiors and their mission, not to open publication.

Case study: The Jesuit cartographers and the problem of "forbidden" world charts

If the BAV was the mind, the Jesuit network was the nervous system. Teachers, astronomers, linguists, and diplomats, Jesuits were superb at turning field notes into durable knowledge. They ran colleges with printing presses; they taught mathematics and astronomy; they advised courts from Europe to Asia; and they compiled maps that were both portable and persuasive.

Here's how the machine worked, in practice:

1) **Intake.** Letters from the field—annual reports, scientific observations, itineraries—arrived at a provincial house. They included latitude estimates (via solar altitudes), distance by day counts, bearings, and local place-name lists.

2) **Reconciliation.** Mathematicians reconciled conflicting reports by weighting observers (the veteran pilot beats the new missionary on coasting detail; the missionary beats the pilot on tribal geography inland). They checked astro-observations against ephemerides taught in Jesuit colleges.

3) **Compilation.** Draft world charts were made in sheets, often by region. Some compilers experimented with projections beyond the familiar plane chart—glimpses of a more mathematical cartography that could keep errors from exploding over long arcs.

4) **Circulation control.** Public versions, stripped of sensitive notations, went to presses and patrons. Internal versions stayed in cabinets: "for study," "for missions," or "for review by superiors."

This is where the phrase "forbidden world charts" has teeth. No tribunal of censors in a dark room stamping "FORBIDDEN" on a

Three Reasons to Withhold a Map

- *It revealed a missionary route that could be interdicted.*
- *It crossed a sensitive treaty meridian (e.g., a line inconvenient to royal claims).*
- *It conflicted with the devotional mappaemundi that still had pastoral use.*

globe; instead, practical but firm circulation fences. A world chart might be withheld because it included royal-sensitive coastlines with better longitudes than were publicly accepted, or because it outlined routes to communities not yet ready for European attention.

Where the anomalies live (and how to read them)

The mainstream model says: early-modern compilers were ingenious synthesizers. They could take bits of accurate coastal knowledge and, by choosing projections cleverly, make the whole look better than the parts. That alone can produce "wonders."

The heterodox model adds: in a few cases, the "whole" looks better than is easily explained by known 15th–16th century surveying—suggesting earlier high-precision prototypes behind the Renaissance compilers. The most discussed examples:

- A 1513 compilation credited by its maker to more ancient sources that placed the Americas and Africa in unusually plausible relative longitudes for so early a date. Whether those longitudes are truly "too good" is debated, but the compiler himself said he worked from older charts. That admission is

The Jesuit Compiler's Toolkit

- *Portable quadrants/astrolabes; eclipse and Jupiter's moons tables (for rough longitudes, where feasible).*
- *Local pilots' rutters and sailing directions blended with inland missionary itineraries.*
- *Drafting conventions taught in colleges: rhumb-line grids, experimental projections, and in some cases, comparative overlays.*

not proof of Ice-Age science—but it is a breadcrumb toward a layered lineage of sources.

- Mid-16th-century compilations (notably those associated with humanists and mapmakers in Europe) that appear, to some readers, to trace parts of Antarctica with less ice. Skeptics argue for mistaken coast-matching or artistic coastal smoothing. Proponents counter that certain bays and promontories align too well to be a coincidence. The most careful versions of this argument tether it to the idea of very old source charts, preserved and recopied, then re-projected by Renaissance scholars who didn't fully grasp what they had.

- Mentions in Renaissance scholarship of an inheritance of charts from "before Alexander," or at least from pre-Hellenistic surveying traditions, echoing a belief that the Alexandrian geographers compiled from older, maritime sources. Again, belief is not proof—but it tells you what some compilers *thought* they were handling.

One more thread matters. If late-Ice-Age coasts were ever mapped by someone (human or otherwise), those maps would show shorelines now underwater. The sea rose by the order of 120 meters from the Last Glacial Maximum into the mid-Holocene. Which means any truly ancient coastal survey would be a poor fit to today's maps unless someone had the patience—and the projection savvy—to "undo" the sea's advance. That is exactly the kind of nerdy, patient activity Renaissance humanists and Jesuit mathematicians excelled at. A few heterodox accounts explicitly argue that certain old-looking coastlines on early-modern charts match continental shelves now drowned, and they connect that to submerged archaeological hints in India, Japan, and elsewhere. Whether those matches survive skeptical scrutiny is still contested. But the *idea* that old shorelines could have been preserved—in text, in myth, in map fragments—isn't absurd.

The politics of holy space

Sacred geography wasn't just about roads to relics. It was also about drawing the world as a theater of providence. That matters for maps because a map can teach.

- A mappaemundi told salvation history: East at the top, Paradise at the far side, Jerusalem at the center.

- A missionary wall map flattened distances to bring Asia "closer" to Europe, a rhetorical choice that supported a universal mission.

- A college atlas grouped "pagan" lands not by proximity but by conversion status, steering students' imaginations.

Blend those with pilot charts and royal master maps, and you have a potent recipe for *controlled* imagination: what to notice, what not to publicize, which routes to sanctify, which to keep "in the files."

Inside a "forbidden" world chart: an anatomy

How to Vet an "Impossible" Map

- *Step 1: Identify projection; re-project carefully before judging fit.*
- *Step 2: Look for composite artifacts (one coastline "slides" relative to its neighbor).*
- *Step 3: Separate artist's smoothing from survey data (check repeated small features).*
- *Step 4: Ask if a known older source could explain the anomaly before invoking the extraordinary.*

Picture a large sheet of laid paper, late 16th or early 17th century, pinned on a Jesuit math master's board. Here's what you might actually see if you were allowed to study it:

- **Projection experiments.** Beyond simple plane charts, you might see an oblique pseudo-conic grid or a rudimentary attempt at a global projection that minimizes distortion along a strategic ocean "corridor." That alone would justify keeping the chart internal while the math matured.

- **Layered notation.** The public layer: coasts, placenames, rhumb lines. The internal layer: small triangles marking magnetic anomalies; annotations like "good water here" or "avoid wintering." Inland, dotted lines follow missionary itineraries to specific villages, with patron-saint abbreviations rather than full names—code to outsiders, obvious to insiders.

- **Sensitive longitudes.** Certain capes line up more plausibly than public charts of the day; perhaps the compiler used lunar distances or Jupiter's moons' timing as a rough longitude control in a few regions. A cautious superior would keep that "in the family" until it was consistent enough not to cause diplomatic mischief.

- **Memory fragments.** Along one high-latitude coast, the delineation is oddly smooth and, in two places, uncannily accurate. A marginal note admits: "coast per antiquior chartae." That's the tell: a fragment copied from an older sheet, re-projected to fit.

Would such a sheet be "forbidden"? In the only sense that matters here: not for general circulation. "Internal use" would be stamped in the practices of the house, not necessarily in ink.

Why a Church librarian would care about Ice-Age arguments

You don't have to accept every bold claim about "impossible maps" to see why a conscientious librarian in Rome would treat certain charts as special. Three reasons:

1. **Provenance sensitivity.** If a sheet appears to inherit, even in part, from older, possibly Byzantine or Islamic sources, assigning credit becomes delicate. Better to study quietly than to provoke quarrels about origins and precedence.

2. **Theological caution.** If marginal notes imply dates or geographies that don't sit comfortably with the learned consensus of the day (say, landforms inconsistent with inherited classical authorities), prudent men would avoid scandal. Quiet study, then careful publication—if at all.

3. **Operational prudence.** If the sheet carries sensitive pilot details, publishing it helps rivals more than the mission.

This is where accounts of preserved ancient cartographic memory dovetail with the logic of Church custodianship. If any institution could have archived "odd" prototypes long enough for a brilliant but careful compiler to re-project them in the Renaissance, it's a network of librarians used to handling dangerous ideas—dangerous not because they are "occult," but because they are politically and pedagogically charged.

The deglaciated ghost: coastlines that don't match today

One reason the "forbidden maps" thesis won't die is that it plugs into a big, simple fact: the sea we know is not the sea of 12,000 years ago. After the last glacial maximum, shorelines marched inland; ancient river mouths drowned; continental shelves turned into fishing grounds. If a memory of old shorelines survived somewhere—in myths, in pilot lore, in a chart copied and recopied—then a Renaissance compiler who unknowingly stitched a piece of that into a modern projection could produce a tantalizing mismatch.

Proponents cite specific cases where early-modern sheets appear to match features of continental margins better than present coasts; they connect those claims to underwater archaeology—horseshoe-shaped structures off India; terrace-like forms near Japan; and other hints of coastal complexity before the sea rose. Even those who are skeptical of the specific matches can grant the premise: a cartographic memory might survive in unexpected places, including libraries that prized and protected "odd" knowledge.

Jesuit field science: how the numbers got on paper

It's easy to forget that Jesuit cartography rode on the back of astronomy classes. Before you can teach theology well, you teach the trivium and quadrivium well; that means math, and that means sky-work. The pipeline looked like this:

- **Astronomy labs in colleges.** Students learned to use quadrants, astrolabes, and later telescopes. Ephemerides were taught as living documents, not recitations.

- **Mission field observations.** Those students, years later in Paraguay, China, or Ethiopia, took solar altitudes for latitude, timed eclipses or occultations when they could, and carried back the rawest kind of data: tables in margins, angles next to place names.

- **Back-room reconciliation.** In Rome (and other centers), those numbers were treated seriously. Outliers were challenged, good series were banked, and better estimates replaced rough ones.

- **Critical humility.** What *wasn't* solid didn't go to press. That is likely one of the reasons internal charts existed at all: to use and improve while keeping the noise out of public view.

Seen this way, "forbidden" charts don't imply conspiracy. They imply quality control—and political awareness—inside a system that took numbers, souls, and sovereignty seriously.

What about the famous enigmas?

Because you'll ask, let's step through the two most discussed map enigmas in the clear.

The 1513 compilation and its "older sources." The compiler explicitly wrote that he used many charts, some older, and that a lost West Indies chart of a famous Genoese mariner had informed him. The sheet shows the relative positions of Africa and South America that some readers find surprisingly plausible for 1513. Others caution: even a lucky composite can look "too good," and projection misreadings can inflate the wow factor. Either way, what matters for our chapter is the compiler's own admission of older sources—a habit echoed by several Renaissance scholars—and the very real tradition of ancient chart preservation through Alexandria and Constantinople into Italy. That makes the Vatican a sensible place to look for prototypes and derivative compilations, whether or not you buy the stronger claims.

A Simple Field Sheet (Reconstruction)

- *Column 1: date/time; Column 2: altitude of sun at noon; Column 3: instrument used; Column 4: remarks ("local name," "chief friendly").*
- *Margin: small sailing sketch with bearings; a cross for a mission house; a dot for a watering place*

Mid-16th-century Antarctic sketches. A handful of maps and globes are said to outline an ice-lighter Antarctica. Advocates insist the fit to subglacial coastlines is too good to dismiss; skeptics argue projection and wishful eye. The cautious reader's position: examine the projection, look for composite artifacts, compare like with like (ice shelf edges move!), and ask whether a known ancient prototype *could* account for a portion of the outline. Keep the bar high—but don't close the library door on the question.

How suppression actually works (and how it doesn't)

Real suppression is boring. It's not bonfires; it's budget lines, cataloging decisions, and "we'll keep that in the cabinet for now." In the world we're studying:

- **Index vs. atlas.** The Index of Prohibited Books dealt with doctrines and dangerous theses. It rarely targeted seamanship data per se. A map would be "suppressed," not usually because of theology, but because it was diplomatic or tactical dynamite—or because it contained mixed content (devotional + operational) better separated before public release.

- **House rules.** Orders had internal rules about what could be shown to patrons or visitors. Sensible, not sinister.

- **Quiet correction.** If an internal chart conflicted with a public engraving, the college might revise next term's engraving quietly rather than advertise the delta.

This quieter model fits the evidence we do have about how church institutions handled delicate material. It also leaves room for why certain charts—especially composites hinting at non-standard sources—would live long lives behind locks.

Reading the room: balanced conclusions

Here's the sober summary you can trust:

- The Vatican world—BAV and related cabinets—had both the motive and the means to control sensitive geographies. That was part of good governance, not necessarily of conspiracy.

- The Jesuit network functioned as a first-rate information engine. Its internal compilations would naturally be better, denser, and sometimes more daring than what went to print.

- Claims that a few Renaissance charts contain echoes of much older, even Ice-Age-era prototype maps are intriguing but not settled. The best versions of those claims recognize the role of projection, composite artifacts, and the likelihood of fragmentary inheritance stewarded by conservative institutions.

- Underwater archaeology and deglaciation studies provide a reasonable *context*—not a proof—for why some old coastal memories might exist. Treat this as a live hypothesis, not a verdict.

In the end, a "forbidden" atlas is less a book than a behavior. It's the habit of separating what edifies from what endangers, what can be

What to Keep in Mind

- *Strong claims need strong, projection-aware matches.*
- *"Secret" often means "private/working," not "occult."*
- *The most plausible path for any ancient prototypes into early-modern charts runs through exactly the institutions we've studied.*

taught broadly from what must be used narrowly, what is settled from what is still a trial balloon. In Rome, across centuries, that habit shaped how the world looked on paper—and how fast new geography reached the eyes of the public.

Somewhere, in a guarded cabinet, there may still be a working sheet with a coastal fragment that doesn't quite match today, annotated in a neat early hand: "from older charts; to be reviewed." Whether that fragment is a troubadour's flourish or a true echo from deep time is a question worth asking well—and a reminder that the most powerful tool in any hidden atlas is not the map at all, but the key that tells you who gets to see it.

Chapter 9

Myths of Lost Continents

If you ask ten researchers what "lost continents" really are, you'll hear ten different answers: a moral parable, a cartographic mistake, a geological impossibility, an allegory, a memory from a world before the seas rose, a colonial fantasy, a spiritual homeland, a hoax, or a hope. This chapter treats them as all of the above—stories with layers. We're going to separate what rocks can tell from what maps suggest, and what myths might be remembering from what later authors invented.

Why "lost continents" refuse to die

People don't cling to ideas for centuries unless those ideas fill a gap. "Atlantis," "Lemuria," and "Mu" fill three:

1. a gap in our sense of scale—civilizations older or wider than we've been taught;

2. a gap in memory—catastrophes that stripped coastlines and archives;

3. a gap in maps—odd survivals whose accuracy seems to overshoot their time.

You'll see all three in play here. Let's start with the best-known of the lot: Atlantis.

Atlantis, Lemuria, and Mu as geographic memories

Atlantis: a dated story with geographic teeth

Plato anchors Atlantis not in the hazy "once upon a time," but at a specific epoch: roughly nine thousand years before Solon—about 9600

BC. It's the most argued-over date in the history of ghost geography, but the detail matters because the end of the last Ice Age sits right there on the timeline. That was the period when global seas rose more than a hundred meters and drowned enormous tracts of the continental shelves. In other words, it's exactly when coastlines we'd recognize today were redrawn, and when anything built along the old coasts would have been bitten off by the sea.

Ancient Tamil sources preserve their own flood chronologies for a now-lost southern land (Kumari Kandam) and even outline three big deluges. When you map those claims against glacial melt pulses, there's a suggestive, if imperfect, resonance between cultural memory and paleogeography: a series of rises rather than a single wall of water. The intriguing part is the numeracy—timetables in commentaries that, when translated into calendar dates, cluster around the very Ice-Age window Plato points to.

Key Date Convergence

- *Plato's drowning of Atlantis: ca. 9600 BC.*
- *Tamil traditions for the "First Sangam" era and catastrophic inundations of southern lands: ~9600 BC, ~7200 BC, ~3500 BP (interpreted from traditional durations).*
- *Scientific backdrop: step-wise sea-level rise as the last Ice Age collapsed, submerging continental shelves worldwide.*
- *(Background reference to Plato's 9600 BC framing: see widely available summaries of Timaeus/Critias.)*

Lemuria: a Victorian land bridge that geology retired

The word "Lemuria" did not drop from temple ceilings; it climbed out of a 19th-century zoology paper. A British biogeographer wondered why lemur fossils show up in Madagascar and India but not Africa, and he proposed—before plate tectonics—that a sunken land once connected them. Later geology made quick work of the idea: continents move, yes, but **whole lost super-continents** in the Indian Ocean that recently sank en masse aren't consistent with plate mechanics. There are indeed drowned microcontinents and plateaus (Zealandia in the Pacific, pieces like Mauritia in the Indian Ocean), but the tidy Victorian bridge called "Lemuria" is obsolete science—kept alive more by esoteric literature than by rocks.

Mu: a 20th-century myth with big claims and thin paper trails

"Mu" is different. It isn't a Victorian land bridge; it's a 20th-century story, given a huge stage by an energetic writer who said he'd seen proof on ancient tablets in India. In his narrative, a mother-continent—"Mu"—stood in the Pacific, home to tens of millions, a source from which Egypt, Mesopotamia, and Mesoamerica all borrowed. Then catastrophe. Then amnesia.

Here's the critical part: the tablets he invoked—written in a secret language of an ancient brotherhood—have never been produced for

What Geology Accepts vs. Rejects

- *Accepts: fragments like Zealandia, Mauritia, volcanic plateaus—but ancient, slow processes; no sudden 10,000-year-ago drop of a whole civilization-bearing continent.*
- *Rejects: a recent, cataclysmic sinking of a single, massive Indian-Ocean "bridge" connecting Madagascar to India.*

modern scholars to inspect and verify. The very *word* "Naacal," tied to a supposed priesthood and script that preserved Mu's story, doesn't originate with him; its earlier, different use related to a completely different migration idea. Across decades, historians and science writers pointed out multiple holes: missing primary artifacts, linguistic leaps, and assertions that treated spiritual literature and rumor as documentary evidence. The result? An influential mythos without the anchor of inspectable, datable sources.

Could ancient maps preserve knowledge of vanished lands?

Here we enter contested territory where enthusiasts see a smoking gun and critics see pareidolia. The question: Do a handful of Renaissance and early-modern maps contain geographical data that seems **older and better** than the cartographers' era should allow?

The Antarctica problem (and why it matters)

Several 16th-century and earlier-style compilations depict a southern land with coastal outlines, mountain chains, and even what look like river estuaries. One famous world map shows Antarctica drawn as if coasts were at least partly ice-free, with rivers draining to open water. Another geographer's Antarctic renders identifiable headlands and islands that modern charts accept—suggesting, at minimum, that the mapmaker handled sources better than contemporaries. And one 18th-century academician produces a southern chart that, remarkably, appears to match the **sub-glacial** outline of the continent as we found it with seismic surveys millennia later, including the partition of East and West Antarctica by a waterway where our modern Transantarctic Mountains stand today. The implication of this family of depictions is straightforward but explosive: either these people made improbable leaps, or they had access to source charts that predated great ice advances, or reflected windows when much of the coast was open water.

Let's be precise about the claims and the caveats:

- **Claimed signal:** Some compiled maps show Antarctica's **coastlines, mountains, and river-like features** in ways that look naturalistic, as if based on observation rather than fancy, plus surprisingly apt longitudes for the Atlantic world that outclass many contemporaries.

- **Method possibility:** In at least one case, the grid imposed by the known compiler seems to be a clumsy overlay on older data drawn on a different projection. When researchers remove the latter grid and reconstruct likely ancient projections, fits improve, and more modern place-matches pop out. That's compatible with **copy-of-a-copy** distortions over centuries.

- **Physical hint:** Cores from the Ross Sea contain layers of fine, well-sorted sediments indicative of river discharge from **ice-free** hinterlands until the mid-Holocene, after which the glacial regime dominates, echoing what those old maps depict as estuaries on that same coast. That doesn't prove ancient mapping, but it knocks down the argument that rivers there are geologically laughable.

- **Conservative counterpoint:** Scholars point out projection errors, coastline scale blow-ups, a Renaissance taste for hypothetical "southern lands," and the danger of **retrofitting** modern coastlines onto old squiggles. They also note that claims about "ice-free Antarctica" often rely on speculative earth-crust-shift models or on assuming long ice-free windows more recent than mainstream glaciology allows. (Even sympathetic reconstructions admit mismatches and patchwork sources.)

If myths remember coastlines, where's the ground truth?

Maps can tease. Sea-level curves can tempt. But archaeology has to weigh in. In the Indian Ocean world, where literary traditions remember a drowned south, teams have identified **submerged structures** off Tamil Nadu. One anomaly sits about five kilometers offshore in ~23 m of water: a large, U- or polygonal-shaped arrangement with wall-like segments and courses of blocks visible beneath marine growth. Side-scan sonar has suggested nearby straight lines that look like long walls. Some divers read it as artificial; others reserve judgment. In any case, if that structure is man-made and sits where it sits due **only** to sea-level rise (not subsidence), it implies a date deeper in the past than orthodox timelines currently allow for large coastal architecture in that region. That's exactly the kind of testable claim "geographic memory" requires.

Case Study: James Churchward and the "Mu Papers"

Let's tackle the most polarizing figure directly. Churchward said a secret cache of ancient tablets—in an extinct language of an esoteric brotherhood—revealed a lost Pacific motherland, "Mu," with tens of millions of inhabitants and a sophisticated science. He claimed he learned the script from an Indian priest, one of only three living holders

What the Old Maps Suggest (and Don't)

- *Suggest: A chain of source charts, copied and re-projected, may preserve older coastal outlines that make more sense than Renaissance guesses.*
- *Don't prove: A global Ice-Age empire, or detailed mid-Pleistocene surveys. The material could be fragmentary, regional, accumulated over many voyages, and later misassembled.*

of the knowledge. He then used that revelation to reinterpret myths globally as "colonies of Mu."

Three things to establish clearly:

1. **Provenance:** No tablets have been produced for open, professional examination. There is no museum case you can visit, no published inscriptions reproducible by independent epigraphers, and no stratified archaeological context.

2. **Terminology drift:** The name "Naacal" was not coined by him and had been used in different speculative contexts before his books. He relocated and repurposed the term to fit his Mu narrative.

3. **Reception:** Contemporary and later scholars critiqued the claims as imaginative but ungrounded—mixing occult motifs with untestable sources; his "colonel" persona and sweeping historical rewrites did not help credibility.

The 23-Meter Question

- *Observed: Offshore anomaly at ~23 m depth; horseshoe/hexagonal sense; laterite-like blocks; possible secondary lower wall; nearby linear features on sonar.*
- *Open issues: Natural vs. artificial? Local subsidence or faulting vs. eustatic sea-level rise? Repeat mapping, cores, and dating are needed.*
- *Why it matters: If purely sea-level driven, such depth implies a very old build date; if tectonic, the date could be younger.*

Antarctica & the 1530s 'Terra Australis: Clean Tracing & Polar Fit

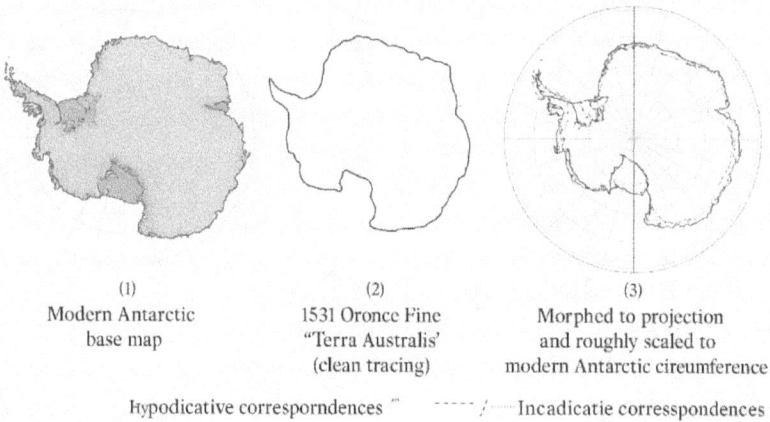

(1)	(2)	(3)
Modern Antarctic base map	1531 Oronce Fine "Terra Australis' (clean tracing)	Morphed to projection and roughly scaled to modern Antarctic circumference

Hypodicative corresporndences ˮ ‑‑‑‑‑ ⌐···· Incadicatie corresspondences

Where does this leave Mu?

As a cultural force, huge. As a historical claim, unproven. That doesn't mean the **Pacific** lacks deep human stories—far from it. It means a modern, all-explaining continent, tied together by unavailable tablets, is not a substitute for dated material culture.

How to Vet a "Revelation Corpus"

- *Is there a physical corpus? (tablets, manuscripts)*
- *Is it accessible? (open to independent scholars)*
- *Is it readable? (clear script, language comparisons)*
- *Is it datable? (radiocarbon, paleography, context)*
- *Is it coherent? (consistent grammar, no anachronisms)*

A middle path: myth as topography, maps as palimpsests

It's tempting to fall into either ditch: "It's all hoax" or "It's all hidden history." A better path is to treat **myths as topographic hints** and **old maps as palimpsests**—layers of copying where some older shorelines may peek through the varnish of later projections.

- In the **Atlantic world**, the 9600 BC Atlantis date and step-wise floods coincide with an era of dramatic coastal change. That is the single best window for a story about a powerful maritime culture losing its lowlands quickly. That doesn't validate every Atlantis-hunt, but it makes the *genre* intelligible.

- In the **Indian Ocean**, the Tamil record of southern lands lost to the sea is at least partly plausible against late-Pleistocene shoreline models. Underwater anomalies off India's southeast coast deserve full, multi-season surveys with cores, photogrammetry, and materials analysis.

- In the **Antarctic debate**, a cluster of old maps really does raise eyebrows. Remove Renaissance gridding, compare on appropriate projections, and credible features emerge more clearly than you'd expect from pure guesswork. At the same time, they don't compel a single grand conclusion about a worldwide Ice-Age civilization; smaller, regionally sophisticated voyaging traditions could also seed a "secret memory of maps."

Practical tests that cut through the fog

If "lost continents" are geographic memories, then the next step isn't arguing harder—it's **measuring better**.

1. **Bathymetry + Cores at Key Sites**
   Identify near-shelf targets where myths cluster (off SE India, the Sunda shelf margins, the Atlantic off

Portugal/Spain/Morocco). Obtain high-resolution bathymetry, run sub-bottom profiling, and core sediments for microfossils and anthropogenic signals (charcoal peaks, industrial minerals, engineered lithics).

2. **Underwater Architecture Protocol**
Before labels like "temple" or "harbor," document with structure-from-motion photogrammetry, test stone type (e.g., laterite vs. bedrock), look for toolmarks, joints, right angles, and repeat units. Then test transport plausibility and anchoring in paleo-shorelines.

3. **Map Forensics**

Build a workflow to:

- o strip later projection grids;

- o test fits on ancient projections;

- o isolate coast fragments and compare statistically (shape metrics) to modern contours without cherry-picking;

- o Publish the pipeline and data for replication. This is how we decide whether we're seeing "coast-like squiggles" or a real signal.

Pulling the threads together

- **Atlantis** may be a dramatization—but its **date** lands in Earth's most aggressive coastal redraft in the last 125,000 years. If you were going to tell a story of a maritime power swallowed "in a day and a night," **that** is when shorelines could change society-level fast. (*For date framing of the Atlantis story, see widely available references.*)

- **Lemuria** began as a scientific guess to explain fossils. Plate tectonics replaced it with a more powerful, testable framework. Fragments exist; a recent, civilization-bearing, Indian-Ocean super-continent does not.

- **Mu** is a modern myth built on inaccessible tablets. That doesn't impugn the urge to look beneath seas; it simply means **Mu** is not evidence.

- **Old maps** truly do contain puzzles. Some southern-continent depictions look unnervingly naturalistic; when re-projected, matches improve. That hints at **deep source libraries** now lost—compilations traveling through Alexandria and Constantinople, later re-gridded in Europe, with errors. It's a plausible path by which fragments of **older coastal knowledge** could survive.

- **Submerged structures** on continental shelves are the decisive arena. Off SE India, one anomaly at 23 m depth remains an enigma—exactly the kind that merits heavy, open-data investigation.

Red Flags vs. Green Flags

- *Red: single-source revelations; closed archives; untestable scripts; one map "proves everything."*
- *Green: converging lines—bathymetry, cores, and multiple map fragments aligning independently; open datasets; reproducible methods.*

What Survives, What Doesn't

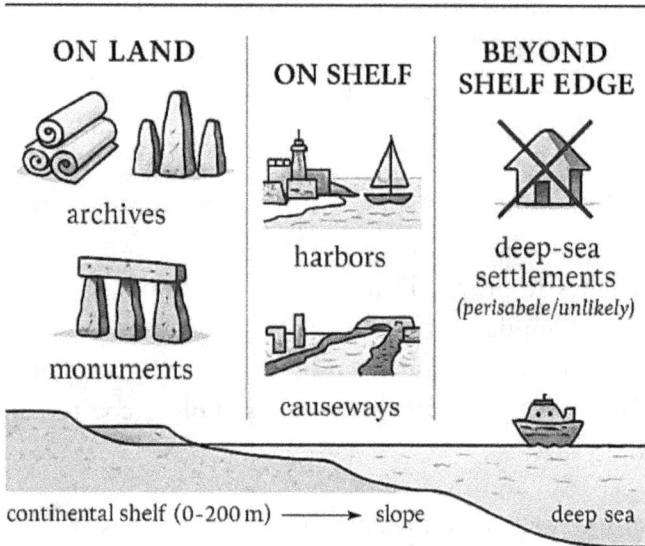

ON LAND	ON SHELF	BEYOND SHELF EDGE
archives	harbors	deep-sea settlements *(perisabele/unlikely)*
monuments	causeways	

continental shelf (0–200 m) ——→ slope   deep sea

A short field guide for you

How to read a "lost continent" claim without getting lost:

- **Ask the rock first.** Is the geology compatible with a recent, sudden sinking? Usually not. But local subsidence, tsunamis, and gradual rise can rearrange coasts in ways legible in cores.

- **Treat maps like evidence, not oracles.** Remove grids, test projections, and share the pipeline. If the fit survives **blind tests**, you're onto something.

- **Weigh myths by the right yardstick.** Their power is not in the literal transcription of events but in **convergent motifs** (multi-stage floods, southern lands, drowned harbors) that point to plausible targets for survey.

- **Demand primary artifacts.** If a claim leans on lost, inaccessible, or single-witness documents, park it. Bring sherds, timbers, mortars, toolmarks, collagen, and dates.

Nothing in this chapter asks you to believe in a globe-spanning Ice-Age super-state. It asks you to **keep your eye on the shelves**—the literal ones (continental shelves) and the archival ones (map rooms). We have cultural memories that plausibly remember coastal drowning. We have old maps that, when you strip the Renaissance scaffolding from them, sometimes look as if they were drafted while the sea's bite mark lay in a different place. We have underwater anomalies at depths that whisper "older than you expect." And we have modern myths—compelling, colorful, and unmoored from hard evidence—that should sharpen our standards, not dull them.

The stakes aren't just academic. Coastlines are where civilizations store wealth, build harbors, and keep archives. When seas rise, the paper trail dissolves. If even a fraction of these "forbidden maps" preserve **real** geographic memory, then the world's drowned edges aren't a romantic backdrop. They're a to-do list.

The Minimal Proof Package

- *A structure under water with: clear anthropogenic geometry; materials not matching local bedrock; datable organics; a paleo-shoreline position that fits sea-level models; independent replication by separate teams.*

Part IV: Redefining Our Past through Forbidden Geography

Chapter 10: Ancient Seafarers and the Case for Prehistoric Navigation

I f you strip away the romance and the movie clichés, navigation is a brutal business. It's problem-solving under pressure: swell patterns that lie, stars that slip behind clouds, a horizon that hides low islands until you're practically on the reef. And yet, long before sextants and chronometers—long before the Age of Discovery wrote its own legend—people were solving that problem at scale. They were crossing open ocean, finding needles in a blue haystack, and doing it with a repeatability that should make any modern mariner nod with respect. This chapter walks you through what the strongest evidence really shows, where the debates still live, and how a handful of "forbidden" clues—old charts, drowned coastlines, and ethnographic memory—force us to widen our timeline for serious seafaring.

Thesis in one line: sophisticated navigation existed millennia before Europe's maritime breakthrough; its fingerprints show up in ancient charts, prehistoric migrations, coastal trade ecologies, and the unmatched precision of Pacific wayfinding.

Evidence of advanced navigation long before the Age of Discovery

What counts as "advanced"?

Two standards separate casual coastal hopping from real blue-water skill:

1. **Open-ocean targeting**—deliberately reaching small, remote islands beyond the visual horizon.

2. **Repeatability**—being able to return along planned routes and stitch voyages together into stable migration and trade networks.

By those standards, several lines of evidence point beyond accidental drift.

- **Prehistoric crossings to island worlds.** People reached Australia by sea more than 40,000 years ago. That migration required multiple water crossings, coordinated groups, and at least practical wayfinding—an early proof that "boats + planning" existed deep in time. In the Holocene, aceramic settlers reached Cyprus, Crete, and other Mediterranean islands; again, no one gets to these without watercraft and intention.

- **The Lapita/Polynesian expansion** demonstrates open-ocean targeting over a million square miles. We'll examine it in depth later, but note here the core: star paths, swell reading, wildlife cues, and canoe design that trades raw speed for stability and endurance over weeks at sea.

- **Cultural toolkits for non-instrument navigation.** Across the Pacific and Indian Oceans, navigators developed codified mental maps—star compasses, seasonal wind calendars, etak/"moving island" frameworks, bird-path heuristics, color-water recognition, cloud-island signatures, and swell refraction patterns. These are algorithms in human memory, not gadgets.

The cartographic wild card: anomalous accuracy on early modern maps

Centuries before modern geodesy and the marine chronometer, a subset of early modern charts appears to embed knowledge that their own eras could not easily have produced. The claim is not that sixteenth-century mapmakers sailed everywhere; rather, that some were **compilers** who drew from older sources. This is where the controversy starts—and where the evidence gets interesting.

- **A 1513 world chart compiled at Constantinople** has long been noted for placing the Atlantic coasts of South America and Africa in strikingly good relative longitudes—something that should not have been achievable in 1513 without a precision time standard. The compiler's own notes say he drew from about twenty earlier maps, some said to be very old.

- **A 1531 world map attributed to a European humanist** shows a large southern land with coastal mountain ranges, estuary-like inlets, and rivers—features that some later readers argued align with sub-ice topography now measured instrumentally. Whether one accepts the identification or not, the point is the **method**: comparative analysis suggested the compiler was working from multiple projections, implying layered source traditions.

- **Eighteenth-century reconstructions** sometimes depict waterways and basins in the southern polar regions that eerily resemble modern subglacial models—again, the argument is about inheritance: was someone copying traces of older geographic intelligence lost to their own time?

Mainstream scholars counter—correctly—that projection errors, coastal misidentifications, wishful redrawing, and the human tendency to see patterns can explain a lot. But even when you discount aggressively, several map features keep landing uncomfortably close to later measurements. That's not proof of a lost civilization; it **is** a nudge to look more carefully at transmission chains, portolan legacies, and the role of large libraries as memory banks that stored older, now-vanished compilations.

Drowned coastlines and the memory problem

The end of the last Ice Age lifted global sea levels by more than 100 meters. That's not trivia; it means that across the continental shelves lie the **old coasts** where late Pleistocene and early Holocene people actually lived. In parts of India's shelf, for example, modeling suggests areas now ~23 m deep were dry land ~11,000 years ago, squarely in the post-glacial transgression window. Reports from offshore Tamil Nadu describe a U-shaped masonry feature at ~23 m depth and other shallow ruins nearer to shore; interpretations vary, but the dating logic runs through sea-level history because diagnostic artefacts are scarce.

None of this proves city-building deep in prehistory; it **does** set the context: the world we can easily excavate is **not** the world people

- *The key claim isn't "people mapped every coast in deep antiquity." It's that **some early modern charts look like re-editions** of information whose origin is older than their compilers—and sometimes older than their own eras could have measured.*
- *Treat these maps as **forensic artefacts of knowledge transmission**: copy upon copy can smear accuracy, but core geometry can survive.*

occupied when shorelines were tens of kilometers seaward. If we're serious about origins of seafaring, we have to be serious about underwater archaeology—and about separating tectonic subsidence, storm scouring, and coastal erosion from sea-level rise when dating submerged structures.

Ancient trade routes across the Pacific and Atlantic

The Pacific: an archipelago of highways

If you want a single case that proves prehistoric people were **systematic ocean navigators**, you study the settlement of Remote Oceania.

The Lapita engine.

Beginning around the second millennium BCE, bearers of a distinctive pottery tradition spread eastward from Near Oceania into the open Pacific—first Melanesia, then out toward Samoa and Tonga, and centuries later into Eastern Polynesia. The archaeology traces obsidian sources, shell ornaments, domesticates, and a ceramic fingerprint over thousands of miles. That is not drift. It's planned expansion guided by **repeatable** route knowledge, island-chain stepping stones, and the ability to sail back the way you came.

The Polynesian crescendo.

Centuries later, Polynesian navigators reached the far corners: Hawai'i, Aotearoa/New Zealand, Rapa Nui/Easter Island, the Societies and Marquesas—a lattice of landfalls that cannot be faked. Their craft—double-hulled voyaging canoes with crab-claw sails—balanced cargo capacity, seaworthiness, and speed. Their wayfinding integrated star altitudes, swells, birds (frigates and terns as "landward messengers"), distant "cloud sitting" over high islands, and the **Te lapa** phenomenon (brief near-surface flashes some navigators associate with island-generated electrical gradients).

Trade ecology, not just migration.

These routes were not one-time shots. Inter-island exchange moved basalt adzes, red feather cloaks, fine mats, canoe timbers, and ideas—religious idioms, genealogies, and navigational lore—across vast distances. Once you have a route and redundancy (multiple voyaging societies), you have a **maritime information network**.

The sweet potato, the chicken, and the contact debate

Botanical and genetic evidence now widely supports **pre-Columbian transfer of the sweet potato** (kumara) from South America to Polynesia. The mechanism—Polynesians reaching South America, South Americans reaching Polynesia, or a complex multi-stop exchange—remains debated, but the simplest navigational interpretation is that **people** met **people** somewhere along that frontier. The broader point: ocean routes are **vectors for biota**, not only for humans. If plants (and possibly chickens, though evidence is mixed by region) made the trip, navigators almost certainly did.

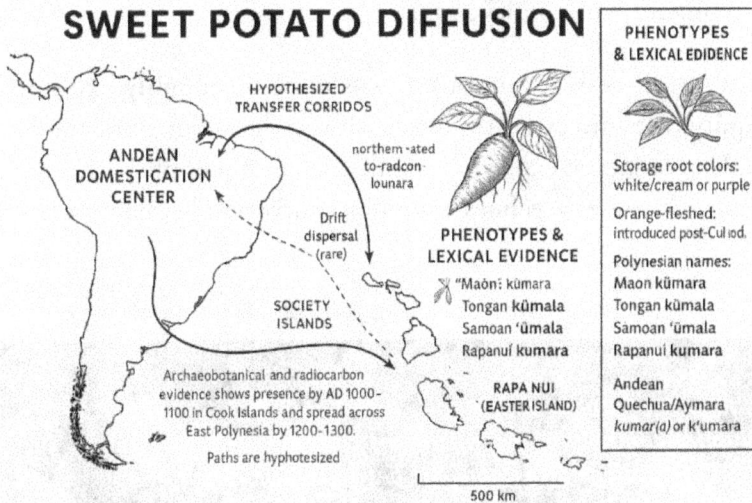

SWEET POTATO DIFFUSION

PHENOTYPES & LEXICAL EDIDENCE

HYPOTHESIZED TRANSFER CORRIDOS

ANDEAN DOMESTICATION CENTER

northern-ated to-radcon lounara

Drift dispersal (rare)

SOCIETY ISLANDS

PHENOTYPES & LEXICAL EVIDENCE

"Maòní kùmara
Tongan kûmala
Samoan 'ûmala
Rapanuí kumara

RAPA NUI (EASTER ISLAND)

Archaeobotanical and radiocarbon evidence shows presence by AD 1000-1100 in Cook Islands and spread across East Polynesia by 1200-1300.

Paths are hyphotesized

Storage root colors: white/cream or purple

Orange-fleshed: introduced post-Cul iod.

Polynesian names:
Maon kûmara
Tongan kûmala
Samoan 'ûmala
Rapanuí kumara

Andean
Quechua/Aymara
kumar(a) or k'umara

500 km

The Atlantic: coastal engines and deep-water rumors

The Atlantic story is more tentative, but it's not empty.

- **Bronze and Iron Age coastal engines.** Mariners in the Mediterranean anchored long trade corridors in metals and luxury goods. Through the Pillars of Hercules, northbound routes harvested Atlantic tin, salt, amber, and fish from the Bay of Biscay to the North Sea. These are not mid-ocean blue-water crossings, but they show how quickly societies scale when a maritime **profit gradient** exists: shipyards, pilotage traditions, seasonality calendars, and port-to-port intelligence.

- **Westbound speculations.** Classical and medieval accounts preserve whispers of islands and lands far to the west. Most likely, these refer to Macaronesian archipelagos (Azores, Madeira, Canaries) or to mirage geography recycled through hearsay. But there are also persistent claims in early-modern charts—fragments on atlases showing Atlantic features and American coasts in ways that imply older sources were in the stew. That claim stands or falls on close technical analysis of the charts themselves, their projections, and their internal geometry—not on legend.

A reasonable synthesis is cautious: **coasts were thoroughly exploited** by antiquity; the **mid-Atlantic** remained mostly rumor until the late medieval/early modern explosion—**unless** a handful of pre-instrument charts really do preserve echoes from older surveys. The map evidence

Pacific case study shows all three pillars of "advanced" navigation long before European instruments:
1. *Targeting small islands far beyond the horizon,*
2. *Round-trip repeatability, and*
3. *Network effects that sustain trade and cultural coherence.*

is not the usual archaeological pottery-and-postholes package; it is **forensic cartography** with all the ambiguity that entails.

Case Study: Polynesian navigation and its startling accuracy

If you want to watch non-instrument navigation at its apex, you talk to a master navigator. The framework below distills what practitioners do under way, how they plan, and why it works.

The canoe is an equation.

A voyaging canoe is a floating compromise:

- **Double hulls** for redundancy and stability in quartering seas.

- **Crossbeams and lashings** are designed to flex without catastrophic failure.

- **Sail geometry** that lets the crew claw upwind in regimes of trade-wind variability.

- **Waterline discipline**—you load for weeks, not hours. That means ballast logic, dry storage, and trim that keeps the bow from burying in a following sea.

The boat is part of the navigator's calculus because a canoe that pounds or hobby-horses changes headings, leeway, and crew performance. Good wayfinding assumes a **predictable platform**.

The star compass: a memory palace on the sky

At its heart, the method is **positional astronomy** without metal. The navigator memorizes:

- **Rising and setting azimuths** of dozens of bright stars for the relevant latitude band.

- **Seasonal drift** of those azimuths as the sun's declination changes.

- **Sidereal "rails"**—parallel tracks defined by star pairs that guide night headings.

A course might be framed as: "Sail until **Altair** rises on the left bow and **Deneb** sets off the right quarter; hold until pre-dawn clouds show anvil form, then follow the swell that wraps from the northeast." That is not poetry. It's a **decision tree**.

The sea speaks: swells, birds, clouds, and color

When the sky goes black or the stars are messy, the sea takes over:

- **Primary swells** run with the trades; **secondary swells** refract around islands, leaving a detectable cross-sea or a beat in the boat's motion. Navigators recognize named patterns—combinations that only occur near certain island clusters.

- **Cloud "sitting"** on high islands is visible beyond the horizon; the color of the undersurface can turn slightly greenish over lagoons.

- **Bird vectors** change by time of day; frigates rarely sleep on water, so dawn/dusk headings can betray roost distance.

- **Water color** and **odor** shifts near large landmasses and river plumes; floating vegetation and insect hits rise with proximity to land.

Dead reckoning without the dead ends

"Etak" (one well-documented framework) treats the canoe as still and **moves the world**—islands advance under memorized star bearings while an internal odometer ticks distance by swell periods and canoe speed. The navigator carries dozens of **etak stages** in memory; each ends when a reference "island" passes through a specified star bearing, then the next stage begins. This flips the Western habit and reduces drift error that accumulates when you imagine yourself moving through an unmarked sea.

Accuracy, you can audit

The acid test is **landfall**. Hitting a high, wide island is easy; hitting a low atoll is not. Yet wayfinders consistently reached targets tiny compared to the search space by using **operational width**—approaching on an axis that maximizes cross-section—and by **expanding boxes**: if the landfall window passed without a sign, they enacted a pre-set search fan using the wind as a fixed reference. Accounts from the first European contact period record indigenous pilots guiding ships across hundreds of miles with confidence; modern revivals have replicated multi-week passages using traditional methods alone, with GPS running only as a silent logbook.

Where "forbidden" geography meets maritime reality

The mainstream and the heterodox positions aren't as far apart as the arguments make them seem when you focus on **method** and **burden of proof.**

- **Mainstream core:** people have been bold mariners far longer than most lay histories admit; the Pacific case is conclusive, the Mediterranean and Indian Ocean cases are strong, and underwater archaeology will likely keep pushing dates and complexity backward.

- **Alternative push:** a set of early-modern charts carries geometry that seems older than their compilers; parts of those charts line up suspiciously well with features measured centuries later; some submerged structures and flood myths point to now-drowned coastal cultures that could have kept maritime knowledge alive through bottlenecks.

The productive middle is simple: **treat maps like artifacts**, test them quantitatively, log where they fail, and be open to multi-century knowledge chains. That approach neither declares a lost global civilization nor dismisses persistent anomalies with a hand-wave.

*Traditional wayfinding is a **complete system**: a tuned vessel, a codified sky/sea lexicon, a planning culture, and a discipline of navigation psychology (sleep patterns, decision audits, and strict adherence to turn-back criteria). It delivers **instrument-grade outcomes** without instruments.*

Putting routes on the water: hypothetical corridors that actually work

This section translates talk into tracklines. These are **practical, meteorology-respecting** corridors a prehistoric mariner could have used.

Pacific stepping stones

- **Bismarcks** → **Solomons** → **Vanuatu** → **Fiji** → **Tonga** → **Samoa** (trade winds east-southeast, counter-currents seasonally exploitable; return legs staged across known inter-island headings).

- **Society Islands hub** fanning to Marquesas and Tuamotus (high islands first, low atolls second, with landfall fans widened by downwind approach).

- **Society** → **Hawai'i** by "northing" into trades' sweet spot before turning west—long, but aligned with summer wind regimes; return by seasonal subtropical westerlies.

Indian Ocean monsoon engine

- **Malay Peninsula** → **Sri Lanka** → **Maldives** → **East Africa**: a seasonal, reversible conveyor—southwest monsoon outbound, northeast monsoon homeward. The monsoon **is** a timetable; prehistoric mariners could attach exchange networks to that clock.

- **Arabian Sea littoral** (Oman–Gujarat–Sindh): coastal stages that can be sailed as a monsoon figure-eight, easily scaled to include island nodes (Lakshadweep, Socotra).

Atlantic prudence

- **Canaries/Madeira → Azores loop** following the Canary Current north then west into the Azores High—entirely doable with Bronze/Iron Age capability *if* a society had the incentive to make the step. The return uses the Portugal Current southbound.

- **West Africa → Northeast Brazil** is a fast downwind ride on the North Equatorial Current—**but** it's a one-way door unless you know the subtropical return conveyor. Viable as a drift scenario, but as a two-way trade route, it demands robust knowledge of the wind belts.

The case for prehistoric navigation—soberly stated

Let's stitch the threads.

1. **Capability:** Multiple world regions show early watercraft, island colonization, and, in the Pacific, an unambiguous demonstration of precise, repeatable open-ocean navigation.

2. **Opportunity:** Post-glacial sea-level rise drowned hundreds of thousands of square kilometers of prime coastal habitat— exactly where early maritime cultures would have flourished and stored boats, gear, and knowledge. We've barely looked.

3. **Transmission:** A subset of early-modern charts likely **compiles** older data. Some features on those charts align with geography their era should not have measured, suggesting an **inheritance** chain through libraries and learned centers that predate the compilers.

4. **Signals in myth and botany:** Flood traditions cluster exactly where sea-level modeling says they should. Cultigens crossed oceanic divides in pre-Columbian timeframes, implying **people** crossed too.

None of that forces a conclusion about a single vanished master-civilization. It **does** compel us to accept that the ancestors of today's maritime peoples were better organized, better connected, and better equipped to solve the sea than the most conservative timelines allow.

How to read the sea the way navigators do

You don't need to be in a canoe to understand the mindset.

- **Think in vectors, not places.** "I'm on a heading with respect to a moving medium; what fixed references do I trust?"

- **Audit your senses.** Is that swell really primary, or is the boat's motion tricking me? Check the timing. Swell periods are a metronome; count them.

- **Normalize for fatigue.** Decision quality drops at 0200. Good navigators build procedures that survive tired brains.

- **Pre-plan failure.** If the sign you expect doesn't appear by X, enact Plan B. Remove pride from the loop.

A final look at the "forbidden" charts

Someone—plural, over time—measured and drew coasts with a level of system that later compilers could inherit. In big libraries, materials migrate, are copied, and get **projected** onto new grids. Errors creep in, but core geometry can persist. When we see early-modern charts whose **relative longitudes** and **coastal morphologies** punch above their weight, we have three choices:

1. Coincidence and misreading;

2. Extraordinary luck coupled with projection artefacts;

3. **Inheritance**—selective survival of older data through centuries of copying.

Option three isn't romantic; it's **mundane**. It says our species has always been good at saving useful scraps and reusing them. On that view, the contentious maps are not "smoking guns"; they're **palimpsests**—eras tracing over eras, with older strokes still visible if you know how to look.

We've established that prehistoric navigation was real, sophisticated, and in some basins truly masterful; that drowned shelves hide significant parts of the story; and that a small family of charts might encode older geographic knowledge. The next step is to test claims of **transoceanic exchange** rigorously: plants, animals, pathogens, technologies, and myths that cross oceans leave signatures. If routes existed, they left **cargo**—not just of goods, but of genes, stories, and symbols.

*Takeaway: treat early-modern charts as **archives**, not single-author products. When a feature consistently matches later measurements across independent copies, ask what **source tradition** could have carried it.*

Chapter 11

The Patterns They Tried to Hide

I deas like world-spanning energy grids, maps of lost continents, and temples aimed at the stars keep coming back because people see patterns—and sometimes those patterns sit on top of hard data that doesn't fit tidy timelines. If you've ever looked at an old chart and spotted a coastline that "shouldn't" be there, or noticed sacred sites that seem to march along suspicious intervals, you know the itch: either the past was more sophisticated than we give it credit for, or we're reading coincidence as design. This chapter treats both possibilities seriously, side-by-side, and shows you exactly where the evidence is strong, where it's thin, and where it points to testable work you (or any team) can do next.

Chapter Promise

- *What keeps myths of global grids, lost continents, and sacred alignments alive is a mix of measurable anomalies, cross-cultural number systems, post-Ice Age sea-level change, and human pattern-recognition.*

- *"Forbidden maps" belong in the main story of human history when they encode accurate geography older than our conventional dates allow—or when they document Ice Age shorelines now underwater.*

- *Case Study: Göbekli Tepe isn't just early; its layout, iconography, and orientation invite astronomical and geographic interpretations that can be checked with clear methods in the field and in the sky.*

Why Myths of Global Grids, Lost Continents, and Sacred Alignments Persist

The core drivers

There are four big reasons these ideas won't die:

1. **Sea-level memory:** Between roughly 17,000 and 5,000 years ago, global sea level rose by more than 100 meters. Coastal cultures saw their shorelines redrawn in slow-motion catastrophe. If any knowledge clung to those vanishing edges—shrines, markers, harbors—its remnants now sit offshore, inviting rediscovery and reinterpretation. Traditions along India's Tamil coast, for example, remember a larger southern land, with legends of cities lost to the sea. Offshore surveys later reported large structures in tens of meters of water, which is exactly where Ice Age inundation would have pushed a truly ancient settlement.

2. **Number patterns across cultures:** Some myth systems obsess over repeating numbers—72, 108, 144, 2160, 4320—numbers that map neatly onto astronomical cycles like the precession of the equinoxes if you track them for long enough. Whether those numbers came from sky-watching or from other organizing schemes, they're present across the ancient world and therefore shape how people look for alignments and "grids."

3. **Cartographic anomalies:** Early modern maps sometimes present coastlines or river mouths "centuries too early," or in forms that look like Ice Age shorelines rather than modern ones. When a 16th-century chart gets a South American feature

right—or seems to sketch an Antarctic form under the ice—researchers reasonably ask what older sources fed it.

4. **Cognitive magnetism:** Humans are wired to spot structure. In sacred landscapes, planners did sometimes lock in cardinal directions, solstitial risings, or round-number spacings that make an intentional "grid" plausible. But our brains also connect dots that don't belong together. Sorting design from coincidence is the job.

The grid idea—what it is and isn't

The grid claim runs like this: certain major monuments or sacred centers fall on longitudes separated by "meaningful" intervals (often multiples of 24 or 72 degrees), suggesting a deliberate, global geodetic framework. Examples commonly cited include the longitudes of Giza, Angkor, a sacred hill at Tiruvannamalai (Arunachala), and a megalithic site in Taiwan, which are described as separated by 48°, 72°, 18°, and 90° steps—numbers that match the precessional sequence and base-3 factors. Proponents argue these are "marker longitudes" in a world grid anchored on a non-Greenwich prime meridian through Giza. Skeptics counter that on a planet full of sites, you can always find some that line up by chance. Both points matter. The fair test is: can we predict new sites using the rule, and do those sites, once located and excavated, independently show planned astronomical or geodetic functions?

How "Forbidden Maps" Fit the Bigger Picture of Human History

What the maps actually show

A few early-modern charts appear to contain data older than the charts themselves. The famous Ottoman chart compiled in 1513 synthesizes multiple sources and captures the Brazilian coastline unusually well— longitudinally correct relative to Africa—and seems to overlay different river depictions (including the Amazon) as though from separate base maps of different dates. On some readings, it even hints at shorelines near the bottom margin that look like an Antarctic coast; others say those are just mis-skewed fragments of South America drawn to fit the parchment. Either way, you're looking at a 16th-century compiler working from older material, and some of that material was surprisingly good.

Another set of charts (world maps attributed to Oronteus Finaeus and later Mercator) depicts a southern continent with mountain ranges, rivers, and inlets placed in ways that—when reprojected and compared to modern subglacial topography—look uncannily like coastlines and drainage if much of West Antarctica were shallow seas. The

---

Precession & the Numbers

- *Observable effect: the background stars drift ~1° every ~72 years at the equinoxes.*
- *Mythic math: 24, 36, 54, 72, 108, 144... recur across stories and ritual counts.*
- *Geodetic hook: advocates claim sacred centers were pegged to longitudes separated by those numbers. The question is whether that pattern predicts new finds, not just explains old ones.*

interpretation here is hotly debated; some argue the resemblance is striking and systematic, others that projection errors and "fill-the-blank-south" traditions explain it. Either way, the exercise is an honest one: identify features, test projections, compare to measured landforms under the ice, and note where it fits and where it fails.

The mainstream counterpoints (and why they're valuable)

Serious historians of cartography have pushed back hard on the "ancient Antarctica" claim, noting that Renaissance mapmakers routinely drew a big southern land to "balance" the globe and that some supposed Antarctic lines can be explained as distorted South American coasts forced to fit the page. They also point out that some early "bedrock match" claims used mid-20th-century seismic interpretations later refined by deeper surveys, meaning any argument should be re-run with the latest geophysics. That's a feature, not a bug: strong claims deserve fresh data and re-tests.

Why these maps still belong in the conversation

Because even after the caution flags, you're left with credible anomalies: accurate longitudes where you don't expect them in 1513; neatly drawn Amazon mouths that imply different river geometries across time; and a persistent southern outline that keeps looking "too good" on certain projections. The sober conclusion isn't "mystery solved," it's "the source pool for these compilations had uneven but sometimes excellent data." That opens two live options: improved late-medieval charting networks we've under-credited—or genuinely ancient surveys in now-lost archives that survived in copybooks. The only way to decide is to do the work: projection analysis, error modeling, and site archaeology, where the maps predict submerged stone.

Lost Continents, Drowned Cities, and the Edge Where Myth Meets Seafloor

The Tamil shelf: legend, bathymetry, and a U-shaped problem

Just off India's southeast coast, divers documented a very large, U-shaped stone structure roughly 5 km offshore at ~23 m depth—deep enough that a purely historic-era ruin is unlikely if simple sea-level rise is the main cause of submergence. Local traditions remember a much larger Dravidian land and cities lost to the sea; inundation modeling shows that between ~12,000 and ~10,000 years ago, that shelf was indeed broader and then shrinking fast. Is the structure natural? Some field teams judged it artificial on balance, citing courses of masonry visible beneath marine growth; others remained cautious pending more excavation. That uncertainty is honest science; the significance is the *location* and *depth*, because it's exactly where a terminal Ice Age shoreline would now be.

Why do these sites keep the "lost continent" talk alive

Because they touch three realities at once: (1) solid sea-level physics, (2) living flood memory in local myth, and (3) actual stone offshore at Ice Age-compatible depths. Even when a given structure proves natural, enough candidates exist to keep the net out. That's not

> *Depth ≠ Date—Use With Care*
>
> - *Depth gives you a candidate date range only if tectonics are minor and subsidence/sedimentation is modeled.*
> - *Rule of thumb on this coast from meltwater modeling places ~20–25 m submergence around 6,000–11,000 BP depending on local factors; do not cite a single "magic" date.*

credulity; that's a rational survey strategy for a drowned shelf.

Sacred Alignments and Global Grids—Separating Signal from Noise

The "base-3 / precession" grid claim mapped

If you root a prime meridian through Giza instead of Greenwich, a set of famous sites fall at longitudes separated by 24°, 48°, 72°, 90°, etc.—exactly the numbers that dominate certain myth counts and match precessional math (1° ≈ 72 years, so 72° maps neatly onto a quarter-precession of sky backdrops). Examples often cited: Angkor at ~72° east of Giza; Tiruvannamalai ~48° east of Giza; a Taiwanese megalithic site almost exactly 90° east and sitting on the Tropic of Cancer. Advocates say that's not random; skeptics say you could pick enough sites to manufacture patterns anywhere. Both views have a point; the filter is *predictive power*. If you can forecast where a yet-unrecognized site should lie and then find it with independent archaeological markers, that's different from post-hoc pattern-spotting.

Case Study — Göbekli Tepe and the Pull of the Sky

What's on the ground

Working Rules for Forbidden Maps

- *Assume mixed sources; parse each coastline as an overlay.*
- *Reproject; test fits against current bathymetry/bed maps.*
- *Use map-derived predictions to guide underwater survey lines.*
- *Accept that some "hits" will be luck—and some "misses" will be noise.*

Set on a hill in southeastern Turkey, Göbekli Tepe is a cluster of mainly circular or oval structures built around tall T-shaped monoliths. The walls are unworked stone; the floors are terrazzo; the enclosures display radiating pillars with two larger T-pillars centered like a pair of guardians. Many pillars carry reliefs of foxes, boars, snakes, birds, scorpions, and abstract signs (including H-like forms and crescents). The quarry pillar sizes can reach toward 50 tons in situ, and estimates suggest hundreds of pillars were raised—astonishing logistical capacity for the tenth millennium BCE. Excavation to date shows no clear habitations; the site looks ritual, a magnet for gatherings.

The alignment question—what we can and can't say yet

- **Orientation:** Several enclosures are roughly north–south with southern entrances, consistent with a community alert to the cardinal frame and seasonal sky paths. That alone doesn't "prove" stellar targeting, but it's the right starting posture if you care about the sky.

- **Iconography:** Pillar 43 (the "vulture/scorpion" relief) and other animal clusters invite celestial readings (constellations encoded as fauna). There's no scholarly consensus, but the

Field Checklist for a Real Geodetic Marker

- *Anchored to a round-number longitude from the grid seed.*
- *Local ritual/iconography codes time cycles (solstice, processional motifs, and base-3 counts).*
- *Physical device (gnomon/obelisk/stele) ties sky to ground at a predictable date/time.*
- *Independent dating shows the marker is truly old, not a modern "alignment."*

motivation to test astronomical mappings is reasonable given how often later cultures did this openly.

- **Ritual timing:** Excavators have suggested periodic gatherings—think seasonally triggered assemblies by lunar/solar cycles. That hypothesis predicts architectural sightlines to heliacal risings/culminations or solstitial azimuths. Those are measurable on-site.

A practical test plan you can run

1. **True north/azimuth survey:** Establish precise azimuths of enclosure axes and entrance sightlines; compare to solstitial/equinox sunrise/sunset azimuths at site latitude in the 10th millennium BCE (precession-adjusted).

2. **Horizon and local topography:** Build a horizon profile; a low southern entrance with benches suggests a staged viewing experience—verify whether any notch or peak frames a particular heliacal rising.

Göbekli Tepe—Hard Facts to Keep Straight

- *Date range: 10th–9th millennium BCE layers for principal enclosures; "oldest monumental stonework yet" is a fair summary.*
- *Build: terrazzo floors, radiating T-pillars, large central twins; quarry block up to ~50 tons.*
- *Use: ritual center; no standard domestic debris; human remains and secondary treatments (excarnation) connected at sister sites.*

3. **Iconography-sky mapping:** For Pillar 43 and other dense panels, test constellation overlays using multiple epoch windows (±500 years around proposed construction) to avoid forcing a single date.

4. **Replication elsewhere:** Apply the same protocol to Nevali Çori and Çayönü (nearby, roughly contemporaneous, with related traits) to check for a coherent regional pattern versus a one-off.

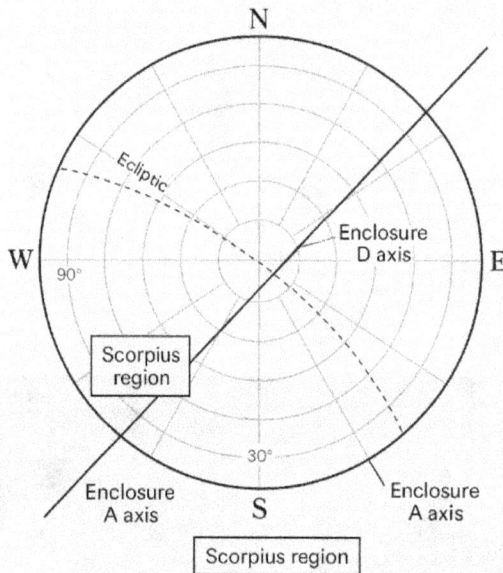

Göbekli Tepe Sky, ca. 9600 BCE
Site latitude 37.22° N

Exploratory overlay—hypothesis generator, not proof.

Why Göbekli Tepe matters to "forbidden geography"

- It's earlier than our old "civilization clocks" and shows heavy stonework and organized labor before classic farming towns fully matured—offering a real candidate locale for sky-ground templating that could later echo elsewhere.

- It demonstrates high-order planning (consistent floor tech, repeated layouts, quarry logistics), which dovetails with the idea that some prehistoric cultures could think in maps—not just of places, but of time cycles.

- It gives us falsifiable alignment targets. Either axes and icons map to specific, recoverable sky moments—or they don't. That's how you sort signal from noise.

What Counts as a "Hidden Pattern"?

The level-headed synthesis

- Yes, patterns exist. Repeating numbers in myth; plausible long-interval site separations; repeated cardinal orientations; map overlays that aren't trivial. Those are real.

- But coincidence is sticky. On a big planet with many monuments, some will land on "pretty" longitudes. Some enclosures will by chance face a notable azimuth. That's why prediction beats retrofitting.

- Underwater is the tiebreaker. If a map predicts masonry at an Ice Age depth off a specific stretch of shelf, and divers find tool-marked blocks under growth that date well before known coastal kingdoms, you've got more than numerology.

How the "forbidden" fits into mainstream history—in one sentence

If even a fraction of the mapping anomalies and offshore structures hold up under modern survey, then practical geography and sky-time tracking were more widespread, and earlier, than our current "firsts" allow; the story of civilization becomes less about a sudden Mesopotamian or Egyptian "switch-on" and more about a deep, incremental knowledge stream that survived the Meltwater Pulses in battered fragments.

 When people talk about "hidden" patterns, they often mean two things at once. First, that gatekeepers dismissed data they shouldn't have (old maps with accurate bits; offshore blocks at Ice Age depths; suspiciously neat longitudes). Second, we lacked the tools to test claims properly. We don't anymore. We can reproject historic charts and compare them to bedrock under kilometers of ice. We can run drone photogrammetry on hilltop sanctuaries and extract azimuths with absurd precision. We can lay sonar grids along drowned shelves where oral traditions say cities stood, and either find walls or find wave-cut terraces—and be honest about which is which.

The world is big. Coincidence is real. So is design. The fastest way to tell them apart is to put the pattern claims on the hook of prediction, roll straight into fieldwork, and let the data decide. If the grid is just in our heads, it will dissolve under measurement. If it's in the ground and the sky, it will sharpen as we map it.

Either way, our past gets more interesting. And that's the point.

What Would Convince a Skeptic?

- A grid-predicted longitude producing a newly discovered, securely dated, pre-Neolithic marker site with clear sky instrumentation (e.g., gnomon sockets, sighting corridors).
- A repeated, statistically unlikely set of separations among independently chosen sacred centers.
- Offshore structures at depths matching modeled sea-level curves and showing unambiguous human tooling.

Chapter 12

The Future of Forbidden Cartography

L et's be direct: the tools now on our desks—and in orbit—are blowing the dust off old charts and revealing ground truth that was literally invisible a decade ago. Lasers rake jungles. Satellites see under the sand. AI aligns warped parchment with modern coastlines. Put together, these methods don't just polish the reputation of "forbidden maps"—they test the claims with numbers, pixels, and ground checks. The result is a new, more rigorous way to read the past.

The New Toolkit of Forbidden Cartography

	LIDAR – canopy-strippecl digital terain model (DTM), hillshade
Buried paleochannel	**SATELLITE SAR** –radar backscatter under desert
	DECLASSIFIED FILM-SCAN – grainy CORONA strip 1600s
	AI GEORECTIFICATION – warped parchment overlaid on modern coitsline grid

Techniques widely used in online archaeology and remote sensing posts (LIDAR Maya/Angkor; SAR Sahara paleochannels; declassified CORONA; AI map recification)

Sensors that strip landscapes bare

Start with LIDAR, a laser altimetry method fired from aircraft. In northern Guatemala, wide-area LIDAR peeled back the canopy across hundreds of square miles and exposed more than 60,000 hidden features—roads, terraces, defensive works, entire urban districts—resizing the Classic-period lowlands from "city-states in the jungle" to an engineered landscape stitched together by causeways. That survey was not a one-off headline; it is now a baseline dataset against which settlement density, logistics, and warfare are being reinterpreted.

The same story played out in mainland Southeast Asia. LIDAR around Angkor revealed a vast, low-density urban complex—grids of ponds, embankments, roads—hidden in plain sight. That work, combining airborne laser scans with careful field verification, rewrote the scale of medieval urbanism in the region and established workflows that have since been reused worldwide.

And then came the Amazon. Aerial LIDAR over southwestern Amazonia documented monumental platforms, causeways, and managed landscapes—"garden cities" whose existence had been argued from chronicles, soils, and small-scale surveys, but finally mapped with crisp elevation data. Follow-ups suggest thousands of earthwork sites and long belts of fortified and ceremonial architecture, reshaping how we talk about "rainforest" and "civilization."

Space-age eyes for Old World problems

"Forbidden cartography" is no longer a metaphor. It's a practical research program that uses LIDAR, radar, declassified spy imagery, and AI to validate or falsify historical map anomalies—region by region, coast by coast.

Orbit has been kind to archaeologists. Declassified CORONA spy imagery (1960s–70s) provides cloud-free, pre-urban sprawl views of landscapes. When you compare those frames to today, erased canals and Bronze-Age mound fields reappear. Analysts stitched thousands of CORONA frames to register settlement patterns across northern Mesopotamia, tracing the lifelines of rivers and roads that later development obscured.

What's new isn't just access; it's automation. Recent work pairs those film scans with machine-learning detectors that can spot the subtle "halos" of tells and the spectral signatures of looting, and then push alerts to teams on the ground. This approach extends beyond "discovery" into real-time heritage protection—a major evolution from the days of hand-loupe inspection.

AI: the palimpsest reader

Take a warped portolan or a 16th-century world map and try to match it to a modern geodesy—you'll hit projection mismatches, hand-drawn distortions, and coastlines that have shifted. AI helps in three ways:

1. **Georectification with uncertainty:** Neural nets trained on control points can nonlinearly "uncrumple" parchment and produce alignment bands that explicitly show where the match is strong and where it is a stretch.

LIDAR is the first method in history that turns "we suspect there's something under the trees" into a map you can test with a tape measure.

2. **Toponym linking:** NLP models match archaic place names to modern equivalents via phonetic, linguistic, and historical "graph walks," flagging when a label likely migrated on copies.

3. **Projection inference:** By testing thousands of candidate projections, algorithms can approximate the grid implied by the drawing—recovering the underlying math a cartographer may have used instinctively.

This matters because the core claims around "forbidden maps" are testable only if we control for projection, deformation, and copying errors. Without that, arguments become opinion pieces. With it, we can say where a coastline corresponds within, say, ±20–50 km—and where it does not.

Where old charts meet new sensors

Several early-modern charts have long been cited for having "too much" geographical knowledge. A famous 1513 Ottoman chart compiled from older sources was rediscovered in an imperial palace in 1929; it shows Atlantic coasts with striking longitudinal relationships for its day and includes notes about using many source maps—some said to be very ancient.

Another 16th-century world map attributed to a learned compiler places a southern continent with individualized coasts, mountain ranges, and river drainages, not as a fantasy blob but as a detailed landmass. Analysts who have studied this sheet in depth argue that the coastal segments look like a compilation of local charts drawn before the ice reached those shores, then re-assembled on a projection grid that does not match the originals—a recipe for misalignments we can now quantify.

A separate mid-century Ottoman map (not the 1513 one) portrays the western hemisphere with a remarkably modern-looking Pacific coastline and even hints at a Bering land bridge—a Pleistocene feature the mapmaker could not have witnessed. The case advanced by technical reviewers is that its western hemisphere must descend from sophisticated source maps on a spherical projection, even as its eastern hemisphere remains conventionally Ptolemaic—an internal inconsistency that points to mixed parentage.

The longitude problem—retested

The hardest historical claim lurking in these anomalies is the whisper of ancient longitudes. Before the 18th-century revolution in timekeeping, longitude was guesswork; even elite navigators were hundreds of miles off on ocean scale. Yet several contested charts sometimes line up better in east-west placement than their era should permit—if, and only if, you assume they were copying long-lost sources. A detailed mid-20th-century analysis made exactly that argument after comparing certain sixteenth-century sheets: where newer compilers leaned on "latest" explorer reports, accuracy dropped; where they leaned on "ancient" sources, relative longitudes improved—suggesting that some predecessors had tools or methods not attributed to historical actors we know. That is a bold claim, but it can now be stress-tested with modern georeferencing and error modeling rather than belief.

Ice, coasts, and dates: what would it mean if some coasts were mapped "too early"?

The pattern across these charts is consistent with a messy copy-chain: accurate coastal "tiles" lifted from older sources, projected (imperfectly), and stitched into later world views. The devil—and the signal—is in the stitching.

The most famous shock point is the southern continent problem. Some early-modern compilations appear to represent segments of a southern land with river systems that would be nonsense under an ice sheet but coherent if coastal mountains were ice-free. Advocates have long argued that such sheets must descend from surveys made in a warmer phase, before ice engulfed the coasts; skeptics counter that we're seeing cartographic pareidolia, projection slippage, and wishful matching. The honest way forward is to keep the question on the table and measure. When researchers re-project one famous 1530s sheet to infer its original grid, and then compare river mouths, capes, and bays to modern coastlines and bathymetry, some sections align far better than chance—others do not. The mixed verdict remains the most defensible one: parts look like recollections of real coasts; parts are composites stretched by later hands.

A 1513 compilation has also been said to echo a southern coastline and sub-ice topography, claims once bolstered by mid-century technical letters noting agreement with seismic profiles gathered in the 20th century. Taken at face value, that's striking; taken with modern caution, it's a prompt to re-run the overlays with today's data and methods, and to separate what survives error bars from what dissolves. Either way, the correct response in 2025 is not to scoff or surrender, but to test.

How modern tech (LIDAR, satellites, AI) is rewriting maps of the past

1. **Jungle to grid**: LIDAR has ended the "empty rainforest" trope. In Mesoamerica, Angkor, and the Amazon, it restores entire engineered landscapes and roads to our maps, shifting population and logistics models from islands of settlement to networks.

2. **Desert to palimpsest**: Historical spy imagery lets us rewind surface damage in arid zones. Pairing CORONA archives with

AI reveals ghost canals, levees, and mounded sites that are otherwise gone. That breathes new life into disputed alignments on old charts—if a medieval coast places a delta farther inland, we can ask whether older channels existed there and whether sediment infill explains the mismatch.

3. **Archive to structured data:** AI can turn shelves of portolans and nautical atlases into a machine-queryable corpus. Automatic toponym extraction, projection inference, and uncertainty-aware warping let us trace copying lineages and see which pieces of a later world map are older "tiles" pasted in.

4. **Fieldwork to probability:** The new flow is digital lead → targeted ground checks → feedback loop. A "possible causeway" in LIDAR becomes a trench and a date. A "suspect river mouth" on a 16th-century sheet becomes a coring transect and a paleo-shoreline model. Each pass either raises or lowers confidence.

What rediscovering forbidden maps means for our understanding of civilization.

First, humility. If laser scans and old charts converge on "more cities, earlier networks," then our baseline narrative—small, isolated, late-blooming—is incomplete. The evidence says humans repeatedly achieved large-scale landscape engineering, long-distance connectivity, and complex planning, often in places we wrote off as inhospitable.

Second, persistence. Early-modern map compilers were not always fantasists. In more than one case, they may have preserved fragments of older geographic knowledge, even while distorting it through their own projections and preconceptions. Treating those sheets as data—noisy data, but data recovers memory that would otherwise be lost.

Third, calibration. Some claims about "too-ancient" precision remain weak under modern scrutiny. The value of this program is not that it confirms every popular mystery, but that it gives us a disciplined way to keep what stands and discard what falls without contempt for either side. Mixed verdicts are progress.

A practical playbook to test "forbidden" cartography claims

- **Assemble the provenance:** Photograph or scan the map at high resolution; record scale bars, compass roses, and any legends about sources or projections.

- **Infer the grid:** Use projection-inference algorithms to test candidates (equidistant polar, portolan-style rhumb frameworks, straight-meridian grids). Keep a ranked list with error margins.

- **Segment the coastlines:** Don't test a whole continent at once. Break the sheet into coastal "tiles" that were likely copied separately.

- **Overlay with uncertainty:** Generate confidence bands against modern coastlines, bathymetry, and paleo-shoreline models. Report fits and misfits, not just a single overlay.

- **Ground-truth anomalies:** If a map shows an unexpected river mouth or island, cross-check with LIDAR (jungle), SAR/optical (desert), CORONA (time-rewind), and, where feasible, cores and trenches.

- **Publish the code and overlays:** Make it easy for skeptics to rerun your steps and for collaborators to build on them.

Archives still matter—maybe more than ever.

The next breakthroughs will combine high-tech scans with low-tech persistence in archives that few people visit. Consider past episodes: an important 1513 chart surfaced because someone opened a neglected cabinet in a palace; other intriguing world maps and portolans were noticed in map rooms only after researchers went hunting for patterns. A major 16th-century sheet showing a detailed southern continent was found and re-examined in just such a trawl, prompting decades of debate. Those stories are reminders that "new" data often arrive as rediscoveries, and that our biggest bottleneck is attention, not bandwidth.

Mainstream vs. alternative: how to keep the conversation honest

Mainstream cautions you'll hear:

- Projection mismatch can create phantom coastlines and misplaced rivers.

- Copying errors and composite sources explain much of the weirdness.

- No confirmed ancient instrument for precise longitude exists in the archaeological record.

Alternative provocations worth testing:

- Some early-modern charts preserve geographic knowledge from long-lost surveys, including coastlines that make best sense in paleoclimatic windows.

- The copy-chain preserved "tiles" with surprisingly good geometry—better than their compilers could independently achieve.

- Non-textual knowledge (maritime itineraries, inherited sailing directions) may encode more precision than we gave credit for.

Your job as an investigator—and our job, collectively—is to move claims out of the realm of reputation and into the realm of reproducible overlays, dated cores, and checkable code. That's not fence-sitting; it's how any field grows up.

A few case paths to pursue next

- **Atlantic "ghost islands"**: Several early-modern charts feature islands (Antilia, Hy-Brasil, etc.) that later mapmakers dropped. Some likely are mirages of copying; others could be bathymetric highs that were exposed during lower sea stands or are misplaced parts of real coasts. Linking these features to bathymetry and paleoshorelines is a tractable project now.

- **Southern-continent segments**: Re-test specific coastal arcs using modern projection inference and overlay against both present shorelines and reconstructions at glacial/early-Holocene sea levels. Publish negative results—they help prune narratives.

- **River mouths and deltas**: A 1513 compilation handles some South American rivers in ways that may reflect different stages of delta growth and sea level. Systematically compare these depictions with stratigraphy and sedimentation models from the same coasts.

- **Underwater shelf mapping**: Where LIDAR or bathymetric lidar is feasible in shallow coastal zones (e.g., shelves off Ireland or Malta), target "islands" and causeways shown on pre-modern sheets. Use side-scan sonar and diver transects to check.

Ethics and provenance—non-negotiables

- **Not harm:** Don't expose sensitive site locations without protection plans.

- **Credit lines:** Archives and local scholars are partners, not sources to strip-mine.

- **Open methods, careful claims:** Share code and overlays; temper headlines.

- **Respect living communities:** Old maps often overlay living heritage landscapes; consult early and often.

What else lies hidden in the archives?

Here's the honest forecast. The future of forbidden cartography looks less like a single smoking-gun map and more like hundreds of careful alignments, cores, and code repositories. It looks like LIDAR teams are partnering with map historians. It looks like a graduate student noticing a weird grid in a forgotten atlas and triggering a year-long projection study. It looks like an Amazon flight plan shaped by a 500-year-old island that might match a shelf ridge. It looks like a CORONA strip that shows a canal we thought was a legend—and a field team that finds its levee.

What else is in those cabinets? Southern-continent fragments we can now measure, not just admire. Coastal "tiles" whose seams betray their

Next steps you can take right now: pick a single early-modern sheet; digitize it well; infer its projection; segment one coastline; publish your overlay and uncertainty bands; invite rebuttals; run a sensor-aided check on the two most interesting mismatches. That's how this field moves from debate to discovery.

older parentage. Marginal notes that quietly say where a captain really got his "own observations." And in the skies and servers: lasers, radar, film, and models waiting for someone to ask the right question.

What does it mean to "redefine our past"

Redefining isn't erasing. It's adding. Adding cities under trees and roads under topsoil. Adding shelf islands to Ice-Age horizons. Adding credible use-cases where early-modern compilers preserved older "tiles" of coastal knowledge—sometimes with startling fidelity, sometimes with obvious seams. Adding a method to our curiosity so that even controversial charts become datasets instead of battlegrounds.

The future of forbidden cartography is practical, collaborative, and testable. And if we do it right, our maps of the past will look less like flat pictures and more like living models—layered, versioned, and ready to change when the next laser sweep or archive note surprises us.

Bonus Section

Forbidden Maps Workbook

Τhis workbook isn't one to sit on a shelf—it gets used. Think of this as your field manual for the ideas in *Forbidden Maps: Ancient Charts, Ley Lines, and the Geographic Mysteries That Redefine Our Past.* You'll get a clean timeline to anchor the big puzzles, step-by-step exercises you can run with modern tools, frank reflection prompts that force you to state what you actually believe, and a resource guide that points you to the places—physical and digital—where the next breakthroughs are waiting.

This workbook is not about winning an argument; it's about sharpening your methods. Where the evidence is strong, we'll help you test it yourself. Where the ground is soft, we'll help you mark the uncertainty and keep going.

> *What makes this workbook different?*
> 1. *It treats you like a co-investigator, not a spectator.*
> 2. *It pairs curiosity with discipline—every cool claim gets a simple test you can run.*
> 3. *It respects both mainstream cautions and alternative hypotheses without hand waving either.*
> 4. *It keeps the tone human, the steps practical, and the stakes clear.*

Part A — Timeline of Mysterious Maps and Their Rediscovery

A strong timeline keeps you from mixing good questions with bad chronology. What follows is a compact, working chronology of the maps and motifs that matter to the "forbidden cartography" conversation, plus key rediscovery moments. Use it as a scaffold; annotate it with your own notes and counterpoints.

Before printing—memory runs on vellum.

- **Late medieval portolans (13th–14th c.):** Practical sea charts optimized for pilotage. Uncannily accurate around the Mediterranean and Black Sea. Their rhumb-line lattices and harbor-heavy coastlines build the habit of "coast first, theory later."

  Investigator's note: These are your baseline for "what careful sailors could achieve without modern longitudes."

- **Classical and Islamic compilations (antiquity–late medieval):** Tables, itineraries, and regional charts survive in fragments and translations. Think of them not as single "master maps" but as libraries of geographic memory that later compilers mined.

  Investigator's note: Custody chains matter—ask how knowledge moved across languages, empires, and script traditions.

The age of compilations—maps that confess they're stitched

- **Early 1500s:** A famous 1513 chart is compiled by a maritime polymath who writes, in plain language, that he synthesized many earlier maps (some said to be very old) and brought them to a common scale. It depicts the Atlantic with striking confidence; debates center on a southern coastal segment that some read as an ice-light Antarctica and others as a distorted extension of South America. *Rediscovery moment (20th c.):* The fragment surfaces in a palace archive, sparking a century of argument.

- 1531: A learned humanist-cartographer engraves a world map whose southern land is drawn with individualized coasts, mountain ranges, and river-like drainage reaching the sea. Projection quirks inflate the landmass, but the hydro-coastal logic keeps the debate alive.

Enlightenment inference—armchairs, but not idle ones

- 18th century (1730s–1760s): A Paris geographer publishes a polar concept map with an inland "frozen sea" and a divided southern land, arguing from watershed logic, iceberg reports, and physical geography rather than eyewitness accounts. Parts of his reconstruction rhyme—loosely but intriguingly—with today's under-ice bed topography.

Industrial and Cold War tools—new eyes on old questions

- Mid-20th century: Seismic profiles and gravity surveys sketch the hidden bedrock of polar regions. Some analysts compare those outlines to Renaissance southern coasts and argue for non-trivial overlaps; others counter with projection artifacts and confirmation bias.

- 1960s–1970s: Declassified film from early spy satellites (high-resolution, pre-sprawl) becomes a time machine for archaeologists; canal lines, mound fields, and ancient road systems reappear on landscapes now altered.

Satellite era—beneath forests, sand, and ice

- 2000s–2020s:
  - Airborne LIDAR pulls engineered landscapes from jungle canopies (Mesoamerica, mainland Southeast Asia, Amazonia).
  - Radar and mass-conservation models reveal Antarctica's under-ice basins, sills, and subglacial

rivers—including long, basin-spanning flows toward the Weddell and Ross sectors.

o Open GIS and digitized archives let anyone overlay historical coastlines on modern basemaps.

Today—what the rediscoveries mean

- The picture that emerges is messier and richer: old charts sometimes preserve older "tiles" of geographic knowledge; modern sensors confirm that past coastlines and waterways can hide under forests, dunes, deltas, and ice. The precise claims vary in strength; the workflow to test them is now accessible to you.

Part B — Exercises: Plotting Ley Lines with Modern Tools

You asked for modern, practical, no-magic methods. Here's a structured set of exercises that takes "ley lines" out of late-night arguments and into repeatable tests. You'll use freely available tools. You'll set tolerances before you draw the first line. You'll run a simple random baseline to see if your "alignment" beats chance. And you'll state your uncertainty like a pro.

What you'll need (free options):

- **Google Earth Pro** (desktop) or **Google Earth Web** for quick experiments.

- **QGIS** (open-source GIS) for rigorous projection control and measurements.

- **A simple spreadsheet** (for logging sites and runs).

- **Optional:** A handheld GPS or phone GPS for field checks; a paper notebook.

Exercise 1 — From Myth to Measurement: Your First Alignment

Objective: Learn to draw, measure, and log one candidate alignment end-to-end.

Steps:

1. **Build your site list (10–30 sites).** Use a trusted gazetteer or official registry. Record name, coordinates (lat/long), date range, and a short note on significance.

Ground Rules Before You Start

1. *Define your site list in advance. Pick a region and set a clear inclusion rule (e.g., "prehistoric stone circles with secure coordinates in [region]," or "temples dated 500 BCE–500 CE within [country]").*
2. *Set your tolerance. For long alignments (>100 km), a tolerance of ±0.2–0.5° azimuth and a corridor of, say, ±1–2 km may be reasonable; scale down corridors for short-distance tests. Declare it up front.*
3. *Pick your path type. Are you testing great-circle (geodesic) lines on a sphere, or rhumb (loxodrome) lines that keep a constant compass bearing? State it and stick to it.*
4. *Run a baseline. Use a random simulation with the same number of sites and the same bounding box to see how often chance yields equal or stronger "alignments."*
5. *No retrospective cherry-picking. If you refine a rule, restart the test and log the change.*

---

2. **Declare the rule.** Example: "Neolithic stone circles with diameters >20 m within [region]. Corridor width: 1 km. Azimuth tolerance at 200 km: ±0.3°."

3. **Draw a great-circle line in Google Earth.**

   o Add a path between Site A and Site B you suspect are aligned.

   o Extend the path beyond both sites to the edges of your region.

4. **Check intersects.** Do any other sites fall within your pre-declared corridor? If yes, log them. If no, that's fine—non-results are data.

5. **Measure azimuths and distances.** In Google Earth, read the bearing; in QGIS (better), reproject to an equal-area projection appropriate to your region, then compute geodesic bearings.

6. **Score the line.** Note: number of sites on the line, total length, site density along the path, and the pithy "why this matters" in one sentence.

7. **Photograph the field reality (optional).** If feasible, visit at least one site. Check inter-visibility (do sightlines exist?), local topography, and any historical evidence of processional routes.

What to write in your log:

- Rule text, corridor width; line type (great-circle or rhumb).

- Site table (name, lat, long, date, source).

- Line summary (bearing, length, sites touched).

- A 1–2 sentence interpretation **and** a 1–2 sentence counter-interpretation.

Questions:

- If this line were **not** real, what would still look convincing about it?

- Does the alignment survive when you slightly perturb endpoints (±0.05°)?

- Are the "hits" clustered near one end, which might indicate local road logic rather than a continental line?

Enter your answers and any required diagrams in the space below.

_____

_____

_____

_____

_____

_____

_____

Inter-visibility Reality Check

If two sites can't see each other (terrain blocks the line of sight) and no evidence exists of signal stations or elevated markers, an astronomical or survey line is less plausible. Use elevation profiles in Google Earth or QGIS to test sightlines.

Exercise 2 — Rhumb vs Great-Circle: Does Your Line Depend on the Model?

Objective: Learn the difference between constant-bearing (rhumb) and shortest-path (great-circle) lines—and see if your alignment is robust.

Steps:

1. Pick one of your best candidate lines from Exercise 1.

2. **In QGIS,** install a plugin or use built-in tools to draw a **rhumb line** between the same endpoints.

3. **Overlay both lines.** In small regions, the difference might be negligible; over long distances, the divergence grows.

4. **Re-score the corridor hits** using the same width.

5. **Interpret:** If your "alignment" only holds for one model and collapses for the other, your conclusion must say so—some ancient surveyors navigated by bearing, not by great-circle math.

Projection Cheat Sheet

- *Great-circle (geodesic): shortest path on a sphere; azimuth changes with longitude.*
- *Rhumb (loxodrome): constant compass bearing; spirals toward the pole on a globe.*
- *Local planar: handy for short distances; can mislead over long arcs.*

Questions:

- Which model better matches the surveying or navigation tools plausibly available to the builders in question?

- Does the cultural context suggest constant bearings (compass-like logic) or astronomical azimuths that vary with latitude?

Enter your answers and any required diagrams in the space below.

_____

_____

_____

_____

_____

_____

_____

_____

_____

_____

_____

_____

_____

Exercise 3 — Beat the Dice: A Simple Random Baseline (Monte Carlo Lite)

Objective: See whether your alignment exceeds what chance would produce under the same constraints.

Steps:

1. **Define your bounding box** for the region.

2. **Keep the same number of sites** as your real list.

3. **Generate 1,000 random site sets** (use a simple script or an online random-point tool; if scripting isn't your thing, 100 runs still teach you a lot).

4. **For each run,** apply your corridor and count "hits" along the same line geometry (or along a line chosen by your rule).

5. **Plot a histogram** of hits; mark where your real alignment sits.

6. **State the result** plainly: "In 1,000 random trials, only 9 produced ≥5 hits; our real line has 7. Under this model, p ≈ 0.009."

Reporting Without Hype

Good statement: "Given our rules and region, this alignment is unlikely under chance alone."
Bad statement: "This proves an ancient global grid."
Better next step: "We'll now test inter-visibility and cultural links."

Questions:

- Does your result depend on a particular corridor width? Try ±50% and re-run.

- Are you using the same **site density** as reality? If your real list clusters near rivers or coasts, reflect that in random draws.

Enter your answers and any required diagrams in the space below.

_____

_____

_____

_____

_____

_____

_____

_____

_____

_____

_____

_____

_____

Exercise 4 — Star Logic or Survey Logic? A Quick Astronomical Cross-Check

Objective: Test whether a claimed alignment corresponds to plausible solar or stellar azimuths for the construction era.

Steps:

1. **Pick a midpoint** along your line and a date range consistent with your sites.

2. **Use a reputable solar/stellar azimuth calculator** to find sunrise/sunset azimuths on solstices/equinoxes, and prominent star risings/settings of the era (accounting for precession if applicable).

3. **Compare your line's azimuth** to those benchmarks (± the tolerance you set at the start).

4. **Interpret culturally.** Does the associated culture have documented solar-stellar symbolism that matches your result?

Questions:

- Could the match be a coincidence? How common is that azimuth across your region's latitude band?

- If the line points to a solstitial sunrise only at one end, is that enough?

Enter your answers and any required diagrams in the space below.

_____

_____

_____

_____

_____

_____

_____

_____

_____

_____

_____

_____

_____

_____

_____

_____

_____

_____

_____

_____

_____

_____

_____

_____

_____

_____

Exercise 5 — Field Truthing: A One-Day On-Site Protocol

Objective: Convert a digital hunch into a field-grade observation.

Checklist for the day:

- Printed map with line/corridor and terrain.

- GPS/phone with an offline basemap.

- Notebook with three columns: *Observation / Measurement / Interpretation (+ alt-explanation)*.

- Camera or phone with geotagging on.

On site:

1. **Stand on the claimed line** near a monument and note whether the line crosses an entrance, axis, or marker.

2. **Check local magnetic declination** if using a compass; don't mistake magnetic for true north.

3. **Photograph sightlines** toward the next site at the line azimuth.

4. **Look for intermediate markers:** barrows, standing stones, knolls that could serve as waypoints.

5. **Leave with a verdict** for that segment: "Supported / Ambiguous / Not supported," and why.

Questions:

- If the builders wanted a precise line here, what did they use in practice—stakes and rope? Sunrise on a known date?

- Are there topographic obstacles that would have forced detours for processions even if the "ideal" line exists on a map?

Enter your answers and any required diagrams in the space below.

_____

_____

_____

_____

_____

_____

_____

_____

_____

_____

Part C — Reflection Prompts: Atlantis and Lost Continents

This section is not about convincing you; it's about asking sharper questions than the internet usually does. Answer in full sentences. Be concrete. "I don't know" is allowed—if you pair it with "this is what would change my mind."

1) What do you mean by "Atlantis"?

- When you use the word, are you referring to a **single Bronze Age culture**, a **composite memory** of several maritime polities, or a **Late Ice Age civilization** far older than mainstream timelines?

_____

_____

_____

_____

_____

_____

_____

_____

_____

- Which **geographical setting** do you consider plausible (Atlantic, Mediterranean, Indian Ocean shelves, elsewhere), and why?

_____

_____

_____

_____

_____

_____

_____

_____

2) What would count as good evidence for a lost maritime culture?

- Name **three independent lines** you'd accept (e.g., underwater architecture with toolmarks + dateable cultural artifacts + coherent settlement pattern).

_____

_____

_____

_____

_____

_____

_____

_____

- Name **two things** you once thought were strong evidence that you now rate as weak.

_____

_____

_____

_____

_____

_____

_____

3) Sea-level change: your working model

- Do you understand the difference between **global averages** and **regional relative sea-level curves?**

_____

_____

_____

_____

_____

_____

_____

_____

_____

- If someone shows you a structure at −23 m, what date range does that invite in your chosen region—and what local factors could shift it?

_____

_____

_____

_____

_____

_____

_____

_____

4) Cartographic memory

- On a spectrum from **hard skeptic** ("all strange maps are projection artifacts") to **strong retention** ("some early maps preserve fragments of Ice Age coasts"), where do you stand **today?**

_____

_____

_____

_____

_____

_____

_____

_____

- What **single test**—if run cleanly—would move your position one notch?

_____

_____

_____

_____

_____

_____

_____

_____

_____

_____

5) Ley lines and intent

- If an alignment survives your **Monte Carlo baseline** and **field checks**, what cultural explanations do you rank highest (astronomy, procession routes, territorial demarcation, geomancy)?

_____

_____

_____

_____

_____

_____

_____

- What **null model** will you keep in mind so you don't overclaim?

_____

_____

_____

_____

_____

_____

_____

_____

6) A statement to future-you

Write a short paragraph beginning, **"Here's what I currently believe about lost continents and why, and here's what would make me change my mind..."**

_____

_____

_____

_____

_____

_____

Part D — Resource Guide: Archives, Digital Map Collections, and Research Projects

This is your launch pad. It's organized by what you actually need to do: look at old maps, compare them to modern geography, test alignments, and explore underwater/coastal data. Every entry is chosen because it's genuinely useful for hands-on work.

1) Digitized Historical Maps

- **National and parliamentary libraries:** Extensive scans of early modern atlases, portolans, and world maps. Look for download options in TIFF or high-quality JPEG and note shelf marks for citation.

- **Specialized map portals:** Aggregators that let you search by date, projection, region, and keyword. These are ideal for building a comparative set (e.g., multiple southern-hemisphere depictions across decades).

- **University collections:** Many host curated exhibitions of portolans and polar maps with technical notes—great for understanding projection quirks and engraving conventions.

How to use them well:

1. Download the **highest-resolution** version.

2. If a **scale bar** exists, capture it—photograph or crop separately for later calibration.

3. Save the **catalog record** (PDF/HTML) in your project folder so you have dates, attributions, and physical dimensions.

GIS Basemaps and Data Layers (to compare old with new)

- **Global coastlines & rivers:** Open datasets (vector) you can bring into QGIS; choose generalized layers for fast work and detailed layers for close fits.

- **Digital Elevation Models (DEM):** Global DEMs for terrain and line-of-sight checks; regional high-resolution DEMs where available.

- **Bathymetry:** Public gridded seafloor depth models for continental shelf studies.

- **Shoreline change & paleo-coastlines:** Reconstructions for the Late Pleistocene–Holocene transition; use them to sanity-check claims about drowned settlements.

- **Polar bed topography:** Antarctica/Greenland bed data (the "land under the ice"), plus ice sheet masks if you're comparing to historical polar depictions.

How to use them well:

- Reproject layers to a **common CRS** before measuring.

- **Style** DEMs with hillshade to make relief and line-of-sight checks intuitive.

- **Clip** global datasets to your region to keep QGIS snappy.

Quick CRS Picks

- *Regional work: local UTM zone (metric, preserves shape locally).*
- *Continental comparisons: Lambert Conformal Conic (mid-lats) or Albers Equal-Area (analysis).*
- *Polar: Polar Stereographic north/south.*

Underwater and Coastal Research Aids

- **Marine charts and Notices to Mariners:** For shelf features, submerged hazards, and historical coastline clues.

- **Side-scan sonar imagery repositories:** Useful for distinguishing geology from structure when available.

- **National coastal survey reports:** Often host PDFs with site sketches, cross-sections, and context photos for drowned harbors and paleo-channels.

- **Tide and sea-level data portals:** For regional relative sea-level curves; use them to anchor dates for submerged features.

Practical tip: Pair bathymetry with **sediment maps** where possible. Natural terraces and wave-cut benches can mimic "steps" that some mistake for architecture.

Software You Can Learn in a Weekend

- **QGIS:** Your main GIS. Learn how to: add layers, set CRS, georeference an image, draw/measure geodesics, run basic statistics, and export figures.

- **Google Earth Pro (desktop):** Fast for first-pass inspections; handy elevation profiles; easy sharing.

Georeferencing a Historical Map (5-Step Sprint)
1. *Choose control points: distinctive capes, river mouths, islands.*
2. *Use the Georeferencer plugin in QGIS.*
3. *Pick a transformation (start simple: thin plate spline or polynomial 1/2).*
4. *Spread control points evenly; avoid clustering.*
5. *Export the georeferenced raster and check residuals; flag areas with high error.*

- **Simple plotting (spreadsheet or Python/R if you're comfortable):** For Monte Carlo histograms and azimuth rose diagrams.

- **Image editors:** For annotating overlays (arrows, labels, confidence bands).

Physical Archives and How to Behave in Them

- **National archives & map rooms:** Request items in advance; bring a list of shelf marks; arrive with pencils, not pens.

- **University special collections:** Often friendlier to researchers without credentials if you communicate clearly.

- **Local museums & private collections:** Sometimes hold unique regional charts; be respectful and clear about your aims.

Archive etiquette essentials:

- Handle with both hands; follow staff instructions; never lean on a map.

- Photograph systematically (top-left to bottom-right) with even light; include a **scale and color card** in at least one frame.Record every inscription, watermark, and scale bar. Those tiny details often unlock projection and source lineage.

What to Photograph (Don't Rely on Memory)

- *Full map (orthogonal).*
- *Details: compass roses, rhumb centers, scale bars, marginal notes.*
- *The entire verso (back) if allowed—ownership marks can matter.*
- *The catalog record on the reading table.*

Part E — Capstone Projects (Optional but Highly Recommended)

Each project pushes you to combine map work, GIS, and critical reasoning. Choose one and finish it. Publish your method and results so others can replicate or challenge you.

Capstone 1 — Re-testing a Southern Continent Segment

Goal: Take a 16th-century southern coastline segment and test it quantitatively against modern polar bed topography.
Deliverables:

- Georeferenced historical segment with control points listed.

- Side-by-side overlay on the modern coastline **and** on bed topography.

- A one-page "fit report" with green (good fit), amber (uncertain), and red (poor fit) segments.

- A plain-language conclusion: "Parts A–B show structural rhyme; parts C–D do not. Here's what that means—and does not mean."

Key questions:

- Does the match survive when you change projections?

- Are you forcing a global fit when local fits make more sense?

Enter your answers and any required diagrams in the space below.

_____

_____

_____

_____

Capstone 2 — A Regional Ley-Line Study That Survives Skepticism

Goal: Build a site list with clean rules, run the corridor and random baseline tests, and write a balanced report.

Deliverables:

- Site list (CSV) with inclusion rule at the top.

- Map figure showing line(s), corridor(s), and hits.

- Monte Carlo histogram with your real result marked.

- Field-truth notes for at least one segment.

- A two-paragraph summary with both an explanation **and** a counter-explanation.

Key questions:

- If the result is borderline, what's your **next** test (inter-visibility, cultural context)?

- Are you willing to publish a **negative** result? (You should. It helps the field.)

Enter your answers and any required diagrams in the space below.

_____

_____

_____

_____

_____

_____

Capstone 3 — Underwater Shelf Reconnaissance (Desk Study)

Goal: Identify a continental shelf target from historical references and map datasets, and assemble a realistic survey brief.
Deliverables:

- A two-page target dossier: bathymetry snapshots, likely currents, seabed type, visibility windows (season/tide), and nearby ports.

- A GIS figure with candidate transects for side-scan or diver survey.

- A risk & ethics paragraph (safety, permissions, environmental care).

Questions:

- Is the "structure" plausibly natural under wave-base and geology?

_____

_____

_____

_____

_____

_____

_____

_____

_____

_____

_____

_____

_____

_____

_____

_____

_____

_____

- Is there a nearby terrestrial counterpart (harbor, quarry) that would strengthen the case?

_____

_____

_____

_____

_____

Part F — Templates and Checklists (Tear-Out Style)

Keep these handy. Re-copy them into your own notebook if you like paper.

Alignment Study One-Pager

- **Region & rule:**

- **Corridor width & line type:**

- **Sites (table):** name / lat / long / date / source

- **Line summary:** bearing / length/sites touched

- **Random baseline:** runs/mean / std dev/percentile of real result

- **Field notes:** inter-visibility / obstacles / cultural context

- **Conclusion (2 sentences):**

- **Alt-explanation (2 sentences):**

Archive Visit One-Pager

- **Repository & call numbers:**

- **Map(s) targeted:**

- **Shots to capture:** full / scale / compass / marginalia / verso / environment

- **Measurements to take:** physical dimensions/scale bar ratios/grid intervals

- **Permissions & constraints:**

- **Post-visit tasks:** georeference/transcribe notes/file naming

Underwater Target One-Pager

- **Location & depth band:**

- **Bathymetry & seabed type:**

- Sea-level curve notes:

- Hazards & logistics:

- Survey method (first pass):

- Criteria for "compelling" vs "interesting":

Enter your answers and any required diagrams in the space below.

_____

_____

_____

_____

_____

_____

_____

_____

_____

_____

_____

_____

_____

Part G — Common Pitfalls (And How to Avoid Them)

1) **Pareidolia on parchment.** Seeing what you want in noisy coastlines.

   Fix: Quantify fit; show misses as well as hits.

2) **Projection amnesia.** Measuring on a pretty picture without checking CRS.

Fix: Always state your projection; reproject before measuring.

3) **Corridor creep.** Widening the corridor until any line looks "good."

Fix: Declare corridor width before you draw; stick to it.

4) **Data cherry-picking.** Adding or removing sites until the line sings.

Fix: Lock inclusion rules up front; restart if you change them.

5) **Ignoring topography.** Proposing a line that runs straight through a mountain wall with no ancient workaround.

Fix: Run elevation profiles; check for passes and saddles.

6) **Ignoring people.** Publishing sensitive site coordinates; trespassing; romanticizing sacred places.

Fix: Ask, inform, anonymize when needed, and behave.

Part H — Intelligent Questions to Carry Into Every Project

These are the questions I ask myself before I claim anything. Make them yours.

- **Signal vs. system:** Is this an isolated wow-moment or part of a pattern that repeats under similar rules elsewhere?

- **Mechanism:** If the alignment or map feature is real, **how** did builders or compilers achieve it with the tools they plausibly had?

- **Transmission:** If a chart looks "too good," what copy-chain could have preserved it across centuries without the original?

- **Date windows:** Do your claims line up with **regional sea-level and climate** phases, not just global averages?

- **Counterfactual:** What observation would make your favorite idea obviously wrong?

- **Peer eyes:** Who will try to break your result, and how can you help them do it fairly (data, code, methods)?

Part I — Quick-Start Walkthroughs (Hands-On, 10 Minutes Each)

These are "I've got an hour tonight" tasks that still move your project forward.

Walkthrough 1: Corridor Heatmap in QGIS

- Add your site points.

- Buffer a candidate line by your corridor width (Create Buffer tool).

- Intersect the buffer with the site layer; count hits.

- Export a clean map with title, scale bar, projection, and your rule text.

Walkthrough 2: Elevation Profile on a Suspected Sightline

- In Google Earth Pro, draw a line between two sites.

- Right-click → "Show Elevation Profile."

- Screenshot and annotate "visible / blocked / skyline feature."

Walkthrough 3: Georeferencing a Portolan Segment

- Load the Georeferencer; add control points at river mouths.

- Choose Thin Plate Spline; check residuals.

- Save and warp; overlay on modern coastline; annotate.

Walkthrough 4: Azimuth Rose

- Pull bearings of axes from three monuments.

- Plot in a rose diagram (spreadsheet or quick script).

- Label clusters (e.g., "solstitial-ish," "cardinal-ish," "other").

Part J — FAQ (The Tough Ones)

Q: Isn't this all just pattern-seeking?

A: It can be. That's why you set rules before you draw lines, use corridors and baselines, and publish misses as well as hits.

Q: Why not wait for experts to settle it?

A: Because experts disagree, and open data lets you contribute responsibly—if you follow good methods.

Q: What if I get a negative result?

A: Celebrate. You just narrowed the search space and saved others' time.

Q: Do I need to code?

A: No. You can do serious work with QGIS, Google Earth, and a spreadsheet. Scripts help, but aren't required.

Q: How do I avoid stepping on cultural toes?

A: Ask. Share your aims. Anonymize sensitive locations. Listen first.

Part K — Closing Letter From Your Future Self

Here's the conversation you'll have with yourself a few months from now if you use this workbook honestly:

"I started with a romantic idea about a perfect line. I ended with a measured result that survived three clean tests and failed one. I wrote both down. That failure taught me more than ten pretty overlays. I learned how projection mangles shapes, how a corridor can be abused, and how often a 'hit' is a coincidence until you beat it with a baseline. I also learned that some old maps carry more signal than I expected—just not always in the place I expected to find it.

I'm not more cynical. I'm calmer. I still want there to be a grand pattern; now I know how to ask whether it's there."

Part L — Your Next Seven Days (A Practical Plan)

- **Day 1:** Install QGIS and Google Earth. Create a project folder; write your first inclusion rule.

- **Day 2:** Build your first site list (10–20 entries).

- **Day 3:** Draw your first line, set a corridor, and log results (even if zero).

- **Day 4:** Run a small Monte Carlo (100 runs). Make the histogram.

Your Pledge (Write and Sign)
"I will state my rules up front, publish my methods, welcome criticism, and change my mind when the evidence warrants it."

- **Day 5:** Field-check one segment (even if it's a city park sightline).

- **Day 6:** Georeference one historical map tile; overlay on a modern basemap.

- **Day 7:** Write a two-sentence conclusion and a two-sentence counter-interpretation. Share with a friend who will argue with you.

Final Word

This workbook doesn't ask you to choose a team. It asks you to measure what you can, admit what you can't, and keep the questions moving. If you use these exercises, ask the uncomfortable questions, and publish your methods as openly as your maps, you'll help push this field out of the fog—whether your favorite mystery survives or not.

And if you ever forget why any of this matters, return to the two ideas

The Researcher's Oath

"I will be serious without being joyless. I will ask for evidence without mocking mystery. I will put my methods where my claims are. I will change my mind when the data change."

that run through every page here: **the world's memory is patchy, and our tools are finally good enough to stitch some of it back together**—carefully, transparently, and with both curiosity and caution switched on.

Conclusion

If this book has done its work, you're finishing with two things in your hands: a bristling bundle of puzzles and a calm, usable method. The puzzles are real—inked onto vellum, carved into stone, submerged on continental shelves, and humming through traditions we too quickly call "myth." The method is simpler than any single mystery: compare, measure, overlay, and admit uncertainty when the data demand it. That stance—disciplined wonder—is the thread that runs through every chapter here. The evidence does not plead for credulity, and it does not deserve ridicule; it asks to be tested. And the more closely we test, the more the record looks like a palimpsest of knowledge: scraps and shards carried across centuries, copied, stretched, hidden, rediscovered, and sometimes misread—yet often stubbornly meaningful.

Consider the through-line of the "southern question," which lit the fuse for this entire inquiry. A 1513 compilation made in Ottoman waters admits, in the compiler's own marginal voice, that it was stitched from earlier charts—some said to be very old—and it depicts Atlantic coasts with a longitudinal confidence that the textbooks tell us belongs to later centuries. A 1531 world map by a European humanist renders a detailed southern land; an Enlightenment geographer later infers interior seas and basins in that same polar realm. Whether one accepts the bolder readings or not, the core pattern is hard to ignore: old maps sometimes organize the far south with a hydro-coastal logic that rhymes with what radar and mass-conservation models now reveal beneath the Antarctic ice—basins, sills, and even coherent subglacial rivers routing toward today's Weddell and Ross sectors. The matches are not 1:1 silhouettes; they are structural echoes that deserve analysis rather than dismissal.

Yet the book never pretends those echoes are proof of Ice-Age surveyors. It insists on the humbler but stronger claim: that early-modern maps often read like re-editions of even older geographic intelligence, sometimes beyond what their own eras could have measured easily. Projection clashes, copy-chain artifacts, and the human appetite for pattern can explain a lot; they cannot, by themselves, explain every case that keeps landing awkwardly close to later measurements. The honest route is the measured one: segment a coastline, choose control points, test fits across projections, and weight misses the same way you weight wins. The map tells you how it was made; believe the marginalia—but verify it.

The portolan thread pushes the same lesson from another direction. Medieval sea charts—rhumb-line webs and harbor-dense coasts—achieve uncanny fidelity across the very waters where accuracy mattered to sailors. The mainstream explanation (cumulative pilotage, compass bearings, and superb workshop craft) is already astonishing enough, but the story doesn't stop there. Some world-scale, portolan-style charts embed geometries and regional accuracies that behave—as soon as you reconstruct their frameworks—as if the draftsman knew more than a single voyage could have supplied. That does not make them Atlantean; it makes them repositories. When you treat them that way, you stop forcing grand secrets onto them and start asking better questions about transmission chains, copybooks, and the libraries that acted as waystations for maritime memory.

The underwater chapters widen the frame again. Late-glacial seas rose drastically; the human coastlines of the terminal Pleistocene now lie across today's continental shelves. That single fact rearranges what "missing evidence" means. If a U-shaped masonry feature sits in ~23 meters of water off India's southeast coast, the first clock you consult is relative sea level, not a tidy dynasty list. If terraces near Yonaguni look eerily architectural, you do not have to decide in the first afternoon whether they are quarries or wave-cut benches; you assemble bathymetry, sediment maps, tool-mark criteria, and context, and you let the geology talk. The point is not to insist that those features are cities; the point is to stop assuming they cannot be because our living shorelines are elsewhere.

Sacred geography, too, becomes practical under this lens. Straight-line romance evaporates when site selection is loose and tolerances are vague. But when you set rules in advance, measure the horizon, and demand cultural context—the "three-leg" test of orientation, horizon, and meaning—some alignments stop being conjurer's tricks and start reading like real instruments: not "energy lines," but social and cognitive lines that tie calendars to corridors, solstices to thresholds, and processions to power. The landscape becomes a circuit because communities wire sky, land, and story together, not because invisible currents glide along ruler marks.

Across these domains—maps, coasts, monuments—the same antagonist keeps appearing, and it is not a villain. It is the knowledge filter: the unavoidable habit of any living discipline to privilege tidy, model-conforming data and to quietly park the rest. That filter is why "forbidden" here has been defined not as illegal or occult, but as the class of evidence that rarely makes the syllabus: maps copied from sources we no longer possess, coasts drowned by sea-rise, alignments ridiculed before they were measured, and archive fragments that survived only because a scribe or a soldier got lucky. Once you name the filter, you can work around it. You can ask not "why did no one know this" but "through which cabinets did this knowledge travel—and what burned along the way?"

The book also sketches how information was sometimes constrained on purpose. Charts circulated in two tiers: public derivatives for display and pedagogy, and internal composites for statecraft, mission, or trade. No hooded tribunal stamped "FORBIDDEN" across a globe; circulation fences did the job just fine. Add to that the pragmatic toolkit of early compilers—quadrants and ephemerides for rough longitudes, pilots' rutters and itineraries, Jesuit drafting conventions—and you begin to see how a map can look "too good, too soon" without magical origins, even as edge cases hint at older prototypes behind the stitching.

If there's a single discipline this book keeps returning to, it is method. "Historical," "scientific," and "mythic" are not rival teams; they are lenses to rotate together. Custody chains and scribal habits tell you what a chart can plausibly be. Geology, bathymetry, statistics, and error bars tell you what a claimed fit actually is. Symbols, rituals, and stories tell you what a landscape meant to its builders, which often predicts where to put your next stake or camera. Keep all three lenses on the bench, and you will make fewer heroic leaps and more cumulative gains. Pair the lenses with four ground rules—no cherry-picking, no single smoking gun, no paranormal escapes, and no ridicule in place of measurement—and even controversial claims become datasets instead of battlegrounds.

Why does this matter? Because if even a minority of the "forbidden" cases survive honest stress tests, we don't erase history; we add to it. Civilizational timelines gain fuzzier, more interesting edges. Navigation's apprenticeship lengthens, hinting that someone, somewhere, used astronomical, geometric, or iterative coastal methods to fix positions better than we expected. Flood narratives and "first-times" look less like fairy tales and more like cultural containers, preserving observations of drowned shelves, sky cycles, and trauma. The cost of ignoring such signals is not merely academic; it blinds us to how knowledge actually propagates—messily, across languages, empires, libraries, monasteries, and markets.

And there is a forward path, not just a stance. The closing chapters and workbook point to a future where rediscovery is a reproducible craft. Digitize the highest-resolution facsimiles you can find. Capture scale bars and marginalia. Reproject everything to a common CRS before measuring. Start with thin-plate splines or low-order polynomials; enumerate your control points; publish residuals and uncertainty bands. Pair bathymetry with sediment maps when evaluating "steps" on a seabed. Post your code and overlays. And—this is crucial—invite rebuttal. These are not ceremonial niceties; they are the only way a contentious field becomes cumulative.

Likewise, treat archives like field sites. Request items in advance; photograph orthogonally and systematically; record every inscription, watermark, and scale bar; and capture the verso and catalog records so your future self—or your critics—can trace provenance and projection. The tiny marks are often the keys: a note that a "new" coast came from an "old" source; a wind-rose center that reveals the geometric backbone of a chart. Courtesy and precision in the reading room pay for themselves when you try to reconstruct a map's skeleton back at your desk.

The same ethic applies on land and sea. For sacred-site work, build dossiers skeptics can learn from: true-north references, horizon profiles, observation logs across seasons, and explicit error budgets. For drowned-coast claims, tie every inference to a local sea-level curve and to geomorphic plausibility. In both cases, resist the urge to force a global fit when the data support only local segments. Score your mismatches as cheerfully as your hits. That is how romantic hypotheses become rigorous questions—and how some of them, surprisingly, survive.

If you want a short list of next steps that would actually move this field, you already have one. Pick a single early-modern sheet, infer its projection, and quantitatively test one southern-hemisphere segment against modern coastlines and bed topography; publish the overlay and a one-page fit report with green/amber/red segments and a plain-language conclusion. Pick one shelf zone where sea-level history and stray reports suggest promise; combine nautical charts, bathymetry, and sediment maps to prioritize survey targets. Pick one sacred precinct; run the "three-leg" test and document horizon events with time-stamped photography across a year. In each case, post your methods and invite replication. That is what it looks like to turn "forbidden" into fieldwork.

And so the book's wager comes into focus. Redefining our past does not mean blowing up foundations; it means widening them—adding cities beneath canopies and roads beneath topsoil, adding shelf islands to Ice-Age horizons, adding credible instances where early-modern compilers preserved older "tiles" with startling fidelity, seams and all. If we live that out—if we keep curiosity yoked to measurement—our maps of the past will stop behaving like flat pictures and start behaving like living models: layered, versioned, and ready to change when the next laser sweep or archival note surprises us. That kind of map does not settle arguments; it makes new work possible.

There will still be whiplash. On some pages, you will feel the awe that we may have underestimated our ancestors yet again; on the next, you will watch a seductive fit collapse under projection math. That is not a failure of the project; it is the project. The gain is twofold. First, we learn to live with the right kinds of uncertainty—the kinds that push us back to the documents, the rocks, the stars, and the sea. Second, we recover a clear, unromantic respect for the ways knowledge actually travels: not as a straight line, but as a tide that advances, retreats, and leaves shells of data on unexpected shores.

So keep the three lenses in your bag. Keep the four rules on your desk. Keep your overlays and code in the open. Then choose one problem small enough to finish and stubborn enough to matter. The world they told you didn't exist is not a conspiracy theory; it is the world we keep rediscovering whenever we scrape soot from the archives, silt from the seabed, and habits from our own thinking. The maps are real. The anomalies are testable. The syllabus will catch up—if we do the work.